Reversed-Phase High-Performance Liquid Chromatography

REVERSED-PHASE HIGH-PERFORMANCE LIQUID CHROMATOGRAPHY

THEORY, PRACTICE, AND BIOMEDICAL APPLICATIONS

ANTE M. KRSTULOVIĆ
Department of Chemistry
Manhattanville College

PHYLLIS R. BROWN
Department of Chemistry
University of Rhode Island

1807 · 175 YEARS OF PUBLISHING · 1982

A Wiley-Interscience Publication

JOHN WILEY & SONS
New York ● Chichester ● Brisbane ● Toronto ● Singapore

Library of Congress Cataloging in Publication Data

Krstulović, Ante M.

  Reversed-phase high-performance liquid chromatography theory, practice, and biomedical applications.

  Includes index.
  1. Liquid chromatography. 2. Biological
chemistry—Technique I. Brown, Phyllis R.
II. Title

| QP519.9.L55K77 | 574.19′285 | 81-15944 |
| ISBN 0-471-05369-4 | | AACR2 |

Printed in the United States of America

10 9 8 7 6 5 4 3

Foreword

Over the past decade HPLC has emerged as the preeminent tool of biochemical analysis. The rapid growth of the technique can be attributed not only to advances in both instrumentation and column technology in liquid chromatography but also to the need for a microanalytical method that has the advantages of gas chromatography and is applicable to nonvolatile substances. Indeed, the high precision of modern liquid chromatographs that embody sophisticated microprocessor technology for control and data handling and the high efficiency of microparticulate packed columns offer a formidable combination for rapid and efficient separation and quantitative analysis of submicrogram quantities at high convenience.

In biomedical applications of HPLC, the use of ion exchangers has been supplanted by that of silica-bonded hydrocarbonaceous stationary phases that are generally referred to as reversed phase chromatography. There are many advantages to this technique that employs a versatile stationary phase and conveniently facilitates the adjustment of the magnitude of chromatographic retention as well as that of selectivity over a wide range by changing eluent composition.

This book is the first dedicated to reversed phase chromatography proper. Its scope is broad, however, as it gives a comprehensive view of the present status of HPLC with particular regard to biomedical applications. Wide interest in HPLC has engendered a great need for books that not only treat theory and instrumentation but also detail the wealth of applications of reversed phase chromatography in the biomedical field. In my view, this volume fills the gap and should serve a wide audience as an introduction, guide, and companion in exploring the wonderful world of HPLC.

CSABA HORVÁTH

New Haven, Connecticut
January 1982

v

Acknowledgments

We would like to thank our many colleagues for their technical assistance and helpful comments and for graciously providing the information used in this book: Drs. Frederic Rabel, James Little, Douglas M. Rosie, Richard A. Hartwick, Sheila Morehouse, Laura Bertani-Dziedzic, and Messrs. Klaus Lohse and Edward P. Kujawa.

We are deeply grateful to Professor Csaba Horváth for taking time from his busy schedule to discuss the various aspects of the book. His expertise and numerous contributions to the field of liquid chromatography have been a great inspiration to us and an invaluable source of information.

Particularly helpful to us has been the skill and patience of Roberta Caldwell in the preparation of the manuscript.

Special thanks go to Dr. Louis Berger for his help in proofreading and the Liquid Chromatography Group at the University of Rhode Island for their long-standing contributions to the field.

ANTE M. KRSTULOVIĆ
PHYLLIS R. BROWN

Contents

Reversed-Phase
High-Performance
Liquid Chromatography

1   Introduction

During the past decade, "high-performance" or "high-pressure liquid chromatography" (HPLC) has become a broadly applicable and valuable analytical tool for scientists in diverse fields. The need for a highly sensitive and simple separation technique has been particularly obvious in biochemical and biomedical research, where progress was often hampered by time-consuming, tedious, or inadequately sensitive or specific methodology. The explosive growth and great popularity of HPLC was catalyzed by advances in column technology and instrumentation. The development of microparticulate packing materials of improved efficiency and stability and the introduction of bonded phases have increased the versatility of the technique and have greatly improved the analyses of multicomponent mixtures.

The reversed-phase mode of HPLC (RPLC), which originated from the fundamental work of Howard and Martin (1), uses a nonpolar stationary phase and a polar mobile phase. It has emerged as the most popular and general HPLC technique. At present, it is estimated that more than 80% of the HPLC separations are performed using this technique, and the field is still expanding rapidly. RPLC is ideally suited for separations of nonpolar and moderately polar compounds. Thus solutes that are not readily separated by normal phase liquid-liquid chromatography (LLC) or liquid-solid chromatography (LSC) can be easily resolved using this technique. In addition, polar and ionic compounds can also be separated by means of secondary equilibria such as ion suppression, ion association (ion pairing), and ligand exchange.

Great operational simplicity, high efficiency, column stability, and the ability to analyze simultaneously a broad spectrum of both closely related and widely different compounds have made this technique the most universal mode of HPLC, and possibly the only mode that new generations will use.

At present, many scientists employing RPLC use the trial-and-error approach in optimizing RPLC separations. Although some chapters of RPLC theory are still being written, we will attempt to present the existing

1

theoretical knowledge scattered in chromatographic literature. Additional topics that are treated include fundamental concepts of HPLC and an up-to-date review of the RPLC applications in the biomedical field. References to the literature and collateral readings are included in each chapter. In our presentation we have adopted a practical, problem-solving approach, since this technique is widely used by researchers in various fields. However, we will try to show that the development of chromatographic strategy, a central step in any analysis, is indeed a logical extension of the existing theoretical principles.

REFERENCE

1. G. A. Howard and A. J. P. Martin, *Biochem. J.*, **46**, 532 (1950).

BIBLIOGRAPHY

Basics of Liquid Chromatography, Spectra-Physics, Santa Clara, CA, 1977.

Bristow, P. A., *LC in Practice*, HETP Publ., 10 Langley Drive, Handforth, Wilmslow, Cheshire, U.K., 1976.

Brown, P. R., *High Pressure Liquid Chromatography: Biochemical and Biomedical Applications*, Academic, New York, 1973.

Deyl, Z., K. Macek, and J. Janak, Eds., *Liquid Column Chromatography*, Elsevier, New York, 1975.

Dixon, P. F., C. H. Gray, C. K. Lim, and M. S. Stoll, Eds., *High Pressure Liquid Chromatography in Clinical Chemistry*, Academic, New York, 1976.

Engelhardt, H., *High Performance Liquid Chromatography*, Springer, Berlin, 1979.

Ettre, L. S., and C. Horváth, "Foundations of Modern Liquid Chromatography," *Anal. Chem.*, **47**, 422A (1975).

Giddings, J. C., *Dynamics of Chromatography*, Dekker, New York, 1965.

Hamilton, R. J., and P. A. Sewell, *Introduction to High Performance Liquid Chromatography*, Chapman & Hall, Wiley-Interscience London, 1978.

Horváth, C., Ed., *High-Performance Liquid Chromatography: Advances and Perspectives*, Academic, New York, 1981.

Johnson, E. L., and R. Stevenson, *Basic Liquid Chromatography*, 2nd ed., Varian Aerograph, Walnut Creek, CA, 1978.

Karger, B. L., L. R. Snyder, and C. Horváth, *An Introduction to Separation Science*, Wiley-Interscience, New York, 1973.

Kirkland, J. J., Ed., *Modern Practice of Liquid Chromatography*, Wiley-Interscience, New York, 1971.

Knox, J. N., J. N. Done, A. T. Fell, M. T. Gilbert, A. Pryde, and R. A. Wall, *High-Performance Liquid Chromatography*, Edinburgh Univ. Press, Edinburgh, 1978.

Parris, N. A., *Instrumental Liquid Chromatography: A Practical Manual*, Elsevier, New York, 1976.

Perry, S. G., R. Amos, and P. I. Brewer, *Practical Liquid Chromatography*, Plenum, New York, 1972.

Pryde, A., and M. T. Gilbert, *Applications of High Performance Liquid Chromatography*, Halsted (Wiley), New York, 1978.

Rajcsanyi, P. M., and E. Rajcsanyi, *High Speed Liquid Chromatography*, Dekker, New York, 1975.

Rivier, J., and R. Burgus, *Biological/Biomedical Applications of Liquid Chromatography*, Dekker, New York, 1978.

Rosset, R., M. Caude, and A. Jardy, *Manual Pratique de Chromatographie en Phase Liquide*, Varian, Orsay, 1975.

Scott, R. P. W., *Contemporary Liquid Chromatography*, Wiley-Interscience, New York, 1976.

Simpson, C. F., Ed., *Practical High Performance Liquid Chromatography*, Heyden, New York, 1976.

Snyder, L. R., and J. J. Kirkland, *Introduction to Modern Liquid Chromatography*, 2nd ed., Wiley-Interscience, New York, 1979.

Tsuji, K., and W. Morozowich, Eds., *GLC and HPLC Determination of Therapeutic Agents*, Part 1, Dekker, New York, 1978.

Unger, K., *Porous Silica: Its Properties and Use as Support in Column Liquid Chromatography*, Elsevier, New York, 1979.

Walker, J. Q., M. T. Jackson, Jr., and J. B. Maynard, *Chromatographic Systems: Maintenance and Troubleshooting*, 2nd ed., Academic, New York, 1977.

II   Basic Theory and Terminology

Chromatography has been classically defined as a separation method in which mixtures are resolved by differential migration of their constituents during passage through a chromatographic column. The separation process is governed by the distribution of substances between two phases: the *mobile phase* (moving phase or carrier) and the *stationary phase*. The extent of the interaction that results between the solute molecules and the molecules of each phase is determined by the physical and chemical properties of the solute molecules in a given environment.

The basic operating forces exerted on solutes can be due to polarity arising from permanent or induced electric fields or London dispersion forces (van der Waals forces) which depend on the relative masses of the solute and solvent molecules. In all forms of chromatography, any variable that can influence the balance of intermolecular forces which are responsible for selective retardation of solute molecules will affect the separation.

After the sample has been carried by the moving phase through the chromatographic bed, the liquid that emerges from the column, the *column effluent*, is a composite of the sample (the *eluite*) and the mobile phase (the *eluent*).

1   BASIC EQUATIONS

The degree of retardation of a particular compound in a mixture is often expressed quantitatively in terms of retention time (t_R). Retention time is defined as the time that elapses from the moment the sample is introduced to the point of maximum concentration of the eluted peak (Fig. 1). More appropriately, retention volumes rather than times are used. The retention volume, V_R, of a given sample component that is equal to the total volume of the mobile phase needed to elute the center of the

4

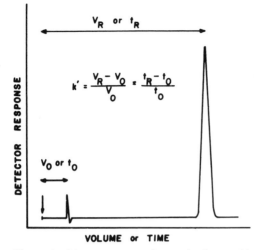

Figure 1 Measurements of capacity factor, k'.

chromatographic band can be calculated from the retention time, t_R, and the volumetric flow rate, F:

$$V_R = t_R F \tag{1}$$

The retention volume is directly related to the fundamental chromatographic parameter, the *distribution* or *partition coefficient*, K, through the following expression:

$$V_R = V_m + KV_s \tag{2}$$

where V_m is the volume of the mobile phase and V_s is the volume of the stationary phase within the column. Compounds that do not interact with the column packing material will be eluted in the *void volume* (dead volume, V_0), which represents the interstitial volume between the particles of packing material and the accessible volume within the particle pores.

Experimental determination of the void volume can sometimes be difficult due to the complexity of separation mechanisms. In RPLC, estimations of the void volume can be obtained using deuterated solvents or solutions of salts under special conditions (1). Exact determinations can be achieved by linearization of plots of homologous series (ln k' vs. carbon number) or from adsorption isotherms.

Retention volume of a sample compound can also be expressed in terms of the elution volume of a nonretained component (V_0). The ratio, which is referred to as the *capacity factor* (retention factor), k', is given by the following expression:

$$k' = \frac{V_R - V_0}{V_0} \tag{3}$$

If the flow rate remains constant during the elution of the sample, expression 3 reduces to:

$$k' = \frac{t_R - t_0}{t_0} \tag{4}$$

The capacity factor is proportional to the free energy change associated with the chromatographic distribution process. It is related to the fundamental chromatographic parameter, the distribution or partition coefficient, K, through the following relationship:

$$k' = \frac{\text{solute mass in stationary phase}}{\text{solute mass in mobile phase}}$$
$$= \frac{\text{solute conc. in stationary phase}}{\text{solute conc. in mobile phase}} \cdot \frac{V_s}{V_m} = K\frac{V_s}{V_m} \tag{5}$$

Thus solute retention is directly related to the thermodynamics of distribution between the two phases.

From expression 5 it is obvious that the retention of a component can be altered by (a) a change in the chemical nature and/or temperature of the two chromatographic phases, and (b) changes in the volumes of the two phases.

The effect of temperature on retention, which is related to the heat of transfer of the solute molecules between the two phases, is given by the van't Hoff equation:

$$d \ln \frac{k'}{dT} = \frac{-\Delta H_{s \to m}}{RT^2} \tag{6}$$

where the nonchromatographic terms have their standard thermodynamic meaning. Since the enthalpies of transfer are relatively small in liquid chromatography (LC), variations in temperature are generally more important in gas chromatography (GC) than in LC. The main advantage in increasing the temperature results from changes in diffusion coefficients and fluid viscosities. Temperature effects will be discussed later.

If the equilibrium concentration of the solute (X) in the mobile phase is plotted versus its concentration in the stationary phase (Y), a series of isotherms are obtained (Fig. 2). Line A represents a linear isotherm and its slope is the distribution coefficient, K. Since K is constant for all values of X, all concentrations in the chromatographic band will move down the column at the same rate, giving rise to a symmetrical elution curve (Fig. 3A).

The concave isotherm (B) occurs when a large sample is placed on the chromatographic column. In this case, the solute-solute forces in the stationary phase are larger than the solute-solvent forces, resulting in increased K values and a nonlinear isotherm. The elution curve has a pronounced leading edge (Fig. 3B), and the retention time of the peak will increase with increasing sample size.

If the forces holding the solute molecules to the surface are only long range due to formation of solute multilayers (in adsorption), the resulting isotherm will have a convex form (C). The elution curve will have a trailing edge, and the retention time of the sample band will decrease with increasing sample size. For analytical separations, the sample size should not exceed the linear capacity of the column, since nonlinearity makes quantitation difficult and prevents the use of retention time data

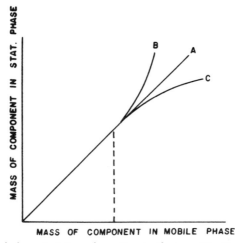

Figure 2 General characteristics of sorption isotherms. (**A**), Linear curve; (**B**), concave curve; (**C**), convex curve.

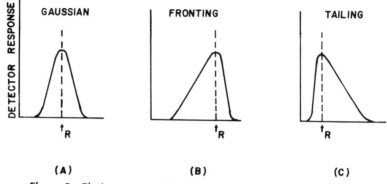

Figure 3   Elution curves: (*A*), Gaussian; (*B*), fronting; (*C*), tailing.

for qualitative analysis. However, this is not the case in preparative work where large sample throughput is sought and column linear capacity is deliberately exceeded.

The values of the capacity factor, k', are characteristic of individual solutes, and the selection of a chromatographic system that will selectively retard components of a mixture is of primary interest. By choosing a proper combination of the mobile and stationary phases, the k' values will be different for each component in a given mixture.

The ratio of the capacity factors, which is referred to as the *selectivity* or *the separation factor*, α, is given by the following expression:

$$\alpha = \frac{k_2'}{k_1'} \tag{7}$$

The capacity factor, k_2', is that of the component with the longer retention time. If α has the value of unity, the two bands are not resolved. If α is larger than 1, the points of maximum concentration of the two peaks are not coincident. However, if the bands are not contained in a small volume of mobile phase, poor separation may result in spite of the favorable α values.

2   SOLUTE-SOLVENT INTERACTIONS

Intermolecular solute and solvent interactions can occur via five possible mechanisms:

1   Dispersion (van der Waals forces or London dispersion forces).
2   Dipole interactions (inductive forces).

3 Hydrogen bonding.

4 Dielectric interactions.

5 Coulombic (electrostatic) interactions.

Dispersion forces operate between molecules with momentary asymmetry of electronic configuration. The temporary dipole in one molecule will, in turn, polarize electrons in adjacent molecules, and the induced dipoles will result in electrostatic attraction between the molecules.

Hydrogen bonding interactions occur between a proton donor and a proton acceptor. The extent of interaction depends on the relative acidity and basicity of the molecules involved.

Dielectric interactions result from electrostatic attraction of solute molecules with a liquid of high dielectric constant. The overall attraction between the solute and solvent molecules is usually a combination of the five types of interactions. The extent of this interaction of a particular solute or solvent is referred to as its "polarity." Solvent strength is directly related to polarity: in normal phase chromatography, the solvent strength increases with solvent polarity, whereas in RPLC, solvent strength decreases with increasing polarity.

2.1 Polarity of Pure Solvents

Polarity of pure solvents can be defined using several solubility indices (1a); the two most common ones are the P' parameter (2,3) and the Hildebrand solubility parameter, δ (1). The P' parameters for different solvents, based on the experimental solubility data of Rohrschneider (2), are given in Table 1. The effect of a change in P' on the k' values in

**Table 1   Properties of Solvents for Use in
Liquid Chromatography**[a]

Solvent	P'
Acetone	5.1
Acetonitrile	5.8
Dioxane	4.8
Ethanol	4.3
Methanol	5.1
i-Propanol	3.9
n-Propanol	4.0
Tetrahydrofuran	4.0
Water	10.2

[a]Reproduced (with modification) from reference 3 with permission.

Table 2 Properties of Chromatographic Solvents^l

Solvent^a	Source^b	UV Cutoff^c	R.I.^{d}	Boiling Point (°C)	Viscosity (cP, 25°C)	P'^e	$\varepsilon^{\circ f}$	Selectivity Group^g	Water Solubility in Solvent^h	Dielectric Constant^i	$P' + 0.25\varepsilon^j$
1. FC-78 (*)^k		210 nm	1.267	50	0.4	<−2	−.25	—		1.88	
FC-75		210	1.276	102	0.8	<−2	−.25	—		1.86	
FC-43		210	1.291	174	2.6	<−2	−.25	—		1.9	
2. Isooctane (*) (2,2,4-tri methylpentane)	LC	197	1.389	99	0.47	0.1	0.01	—	0.011	1.94	0.1
3. n-Heptane (*)	LC	195	1.385	98	0.40	0.2	0.01	—	0.010	1.92	0.5
4. n-Hexane (*)	LC	190	1.372	69	0.30	0.1	0.01	—	0.010	1.88	0.5
5. n-Pentane (**)	LC	195	1.355	36	0.22	0.0	0.00	—	0.010	1.84	0.5
6. Cyclohexane	LC	200	1.423	81	0.90	−0.2	0.04	—	0.012	2.02	0.5
7. Cyclopentane (*)	LC	200	1.404	49	0.42	−0.2	0.05	—	0.014	1.97	0.6
8. 1-Chlorobutane (*)	LC	220	1.400	78	0.42	1.0	0.26	VIa		7.4	2.8
9. Carbon disulfide		380	1.624	46	0.34	0.3	0.15	—	0.005	2.64	1.7
10. 2-Chloropropane (**)		230	1.375	36	0.30	1.2	0.29	VIa		9.82	3.7
11. Carbon tetrachloride	LC	265	1.457	77	0.90	1.6	0.18	—	0.008	2.24	2.3
12. n-Butyl ether		220	1.397	142	0.64	2.1	0.25	I	0.19	2.8	2.4
13. Triethylamine			1.398	89	0.36	1.9	0.54	I		2.4	2.4
14. Bromoethane (*)			1.421	38	0.38	2.0	0.35	VIa		9.4	4.3
15. i-Propyl ether (*)		220	1.365	68	0.38	2.4	0.28	I	0.62	3.9	3.2
16. Toluene	LC	285	1.494	110	0.55	2.4	0.29	VII	0.046	2.4	2.9
17. p-Xylene		290	1.493	138	0.60	2.5	0.26	VII		2.3	3.0

No.	Solvent											
18.	Chlorobenzene			1.521	132	0.75	2.7	0.30	VII		5.6	4.1
19.	Bromobenzene			1.557	156	1.04	2.7	0.32	VII		5.4	4.1
20.	Iodobenzene						2.8	0.35	VII			3.7
21.	Phenyl ether			1.580	258	3.3	3.4		VII		3.7	4.9
22.	Phenetole			1.505	170	1.14	3.3		VII		4.2	4.0
23.	Ethyl ether (**)	LC	218	1.350	35	0.24	2.8	0.38	I	1.3	4.3	
24.	Benzene	LC	280	1.498	80	0.60	2.7	0.32	VII	0.058	2.3	3.6
25.	Tricresyl phosphate								VIa			
26.	Ethyl iodide			1.510	72	0.57	2.2		VIa		7.8	4.2
27.	n-Octanol		205	1.427	195	7.3	3.4	0.5	II	3.9	10.3	5.8
28.	Fluorobenzene			1.46	85	0.55	3.1		VII		5.4	4.6
29.	Benzylether			1.538	288	4.5	4.1		VII			
30.	Methylene chloride (**)	LC	233	1.421	40	0.41	3.1	0.42	V	0.17	8.9	5.6
31.	Anisole			1.514	154	0.9	3.8		VII		4.3	4.6
32.	i-Pentanol			1.405	130	3.5	3.7	0.61	II	9.2	14.7	7.3
33.	1,2-Dichloroethane	LC	228	1.442	83	0.78	3.5	0.44	V	0.16	10.4	6.3
34.	t-Butanol	LC		1.385	82	3.6	4.1	0.7	II	miscible	12.5	
35.	n-Butanol	LC	210	1.397	118	2.6	3.9	0.7	II	20.1	17.5	8.3
36.	n-Propanol	LC	240	1.385	97	1.9	4.0	0.82	II	miscible	20.3	
37.	Tetrahydrofuran (*)	LC	212	1.405	66	0.46	4.0	0.57	III	miscible	7.6	
38.	Propylamine (*)			1.385	48	0.35	4.2		I	miscible	5.3	
39.	Ethylacetate (*)	LC	256	1.370	77	0.43	4.4	0.58	VIa	9.8	6.0	5.8
40.	i-Propanol	LC	205	1.384	82	1.9	3.9	0.82	II	miscible	20.3	
41.	Chloroform (*)	LC	245	1.443	61	0.53	4.1	0.40	VIII	0.072	4.8	5.6
42.	Acetophenone			1.532	202	1.64	4.8		VIa		17.4	8.7
43.	Methylethyl ketone (*)	LC	329	1.376	80	0.38	4.7	0.51	VIa	23.4	18.5	9.1
44.	Cyclohexanone			1.450	156	2.0	4.7		VIa		18.3	9.1

Table 2 (Continued)

Solvent[a]	Source[b]	UV Cutoff[c]	R.I.[d]	Boiling Point (°C)	Viscosity (cP, 25°C)	P'[e]	ε^{o}[f]	Selectivity Group[g]	Water Solubility in Solvent[h]	Dielectric Constant[i]	$P' + 0.25\,\varepsilon$[j]
45. Nitrobenzene			1.550	211	1.8	4.4		VII		34.8	13.2
46. Benzonitrile	LC	215	1.526	191	1.2	4.8		VIb		25.2	10.9
47. Dioxane	LC	265	1.420	101	1.2	4.8	0.56	VIa	miscible	2.2	
48. Tetramethyl urea			1.449	175		6.0		III		23.0	10.7
49. Quinoline			1.625	237	3.4	5.0		III		9.0	7.4
50. Pyridine		380	1.507	115	0.88	5.3	0.71	III	miscible	12.4	
51. Nitroethane			1.390	114	0.64	5.2	0.6	III	0.9		
52. Acetone (*)	LC	330	1.356	56	0.30	5.1	0.56	VII	miscible		
53. Benzyl alcohol			1.538	205	5.5	5.7		VIa	miscible	13.1	8.8
54. Tetramethyl guanidine						6.1		I			
55. Methoxyethanol	LC	210	1.400	125	1.60	5.5		III	miscible	19.9	
56. Tris(cyanoethoxy)propane	GC					6.6		VIb			
57. Propylene carbonate	LC					6.1		VIb			
58. Ethanol	LC	210	1.359	78	1.08	4.3	0.88	II	miscible	24.6	
59. Oxydipropionitrile	GC					6.8		VIb			
60. Aniline			1.584	184	3.77	6.3	0.62	VIb		6.9	8.1
61. Acetic acid	LC	190	1.370	118	1.1	6.0		IV	miscible	6.2	
62. Acetonitrile (*)	LC	190	1.341	82	0.34	5.8	0.65	VIb	miscible	37.5	
63. N,N-dimethylaceta- mide	LC	268	1.436	166	0.78	6.5		III	miscible	37.8	
64. Dimethylformamide	LC	268	1.428	153	0.80	6.4		III		36.7	
65. Dimethylsulfoxide	LC	268	1.477	189	2.00	7.2	0.75	III	miscible	4.7	

No.	Solvent		UV	n	bp	η			M	%w	ε	
66.	N-methyl-2-pyrrolidone	LC	285	1.468	202	1.67	6.7		III		32	10.0
67.	Hexamethyl phosphoric acid triamide	LC		1.457	233	3	7.4		I	miscible	30	
68.	Methanol (*)		205	1.326	65	0.54	5.1	0.95	II		32.7	
69.	Nitromethane		380	1.380	101	0.61	6.0	0.64	VII	2.1		
70.	m-Cresol			1.540	202	14	7.4		VIII		11.8	
71.	N-methylformamide			1.447	182	1.65	6.0		III	miscible	182	
72.	Ethylene glycol			1.431	182	16.5	6.9	1.11	IV	miscible	37.7	
73.	Formamide			1.447	210	3.3	9.6		IV	miscible	111	
74.	Water	LC		1.333	100	0.89	10.2		VIII	miscible	80	

Source: From (1a).

[a] (*) indicates preferred LC solvent of low viscosity (≤0.5 cP), yet with convenient boiling point (>45°); (**) indicates very low viscosity *and* low boiling point solvent.

[b] LC indicates that solvent can be purchased specifically for use in LC from one of following suppliers: Burdick & Jackson, Baker Chemical, Mallinkrodt Chemical, Fisher Scientific, Waters Associates, Manufacturing Chemists, Inc. (GC indicates the solvent is used as a gas chromatography stationary phase and can be purchased from companies selling GC columns and phases (these solvents are used as stationary phases in liquid-liquid LC with mechanically held phase).

[c] Approximate wavelength, below which solvent is opaque.

[d] Refractive index at 25°C.

[e] Solvent polarity parameter (3).

[f] Solvent strength parameter for LSC on alumina (5).

[g] See reference 4, p. 262.

[h] %w of water dissolving in given solvent at 20°C; of interest in LSC.

[i] At 20°C.

[j] Function of P' and dielectric constant that is proportional to solvent strength in ion-pair chromatography;

[k] Fluorochemical solvents available from Minnesota Mining & Mfg. Co.

[l] Reproduced from reference 27 with permission.

RPLC can be expressed by the following relationship (4):

$$\frac{k_2'}{k_1'} = 10^{(P_2' - P_1')/2}$$ (8)

Generally, a change in P' by two units will cause an approximate change in k' values by a factor of 10.

A more common index of solvent polarity is the Hildebrand solubility parameter, δ. A general relationship of δ, P', and the adsorption solvent strength parameter, $\varepsilon°$, is shown in Fig. 4. From this plot, the P' values of solvents that are not listed in Table 1 can be determined. Disagreement between the actual P' values and those predicted from the δ values occurs with basic solvents that do not exhibit donor properties (diethyl ether, triethylamine, etc.). Thus the predicted P' values are higher by approximately 1.5 units.

2.2 Solvent Mixtures

Although it is evident by inspection of Table 2 that a change in solvent can result in a change in k' values of the order of 10^5, a more convenient way of optimizing the separation is by using solvent mixtures. The polarity

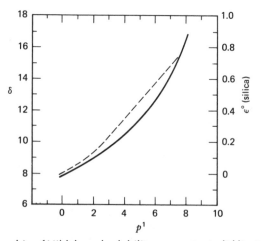

Figure 4 Relationship of Hildebrand solubility parameter (solid line) and adsorption solvent strength parameter (dashed line) to solvent polarity parameter, p. Reproduced from reference 27 with permission.

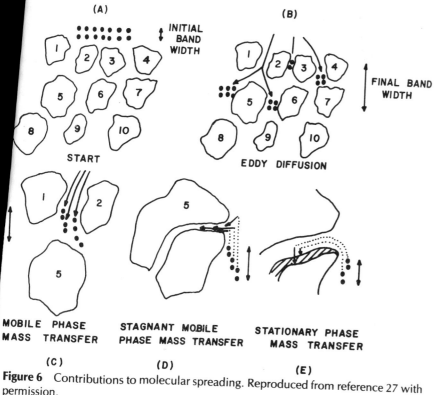

(A)

•••••••• ↕ INITIAL
•••••••• BAND
WIDTH

1 2 3 4

5 6 7

8 9 10

START

(B)

1 2 3 4

5 6 7

8 9 10

FINAL BAND
WIDTH

EDDY DIFFUSION

1 2 5

5

MOBILE PHASE
MASS TRANSFER

STAGNANT MOBILE
PHASE MASS TRANSFER

STATIONARY PHASE
MASS TRANSFER

(C) (D) (E)

Figure 6 Contributions to molecular spreading. Reproduced from reference 27 with permission.

mobile phase flow. The term becomes significant only under stopped-flow conditions and at low flow rates. Because of the low solute diffusivity in liquids, this term is usually negligible in HPLC.

3.1.3 Mobile Phase Mass Transfer

Mobile phase mass transfer can be subdivided into two terms: the moving mobile phase mass transfer (Fig. 6C) and the stagnant mobile phase mass transfer (Fig. 6D). The moving mobile phase mass transfer results from the differences in the mobile phase velocity within a single streamline: solute molecules in the midstream will move faster than those close to the particle surface. The band broadening due to the stagnant mobile phase mass transfer results from the existence of stagnant mobile phase in the intraparticle void volume. Since solute molecules diffuse through

(P') values of a mixture of two pure solvents can be calculated using the following formula (4):

$$P' = \phi_1 P'_1 + \phi_2 P'_2 \tag{9}$$

where ϕ_1 and ϕ_2 are the volume fractions of each solvent in a mixture, and P'_1 and P'_2 are the polarity indices of the pure solvents.

By selecting an appropriate solvent pair for a particular separation, the k' values for the sample component will be in the range of 2 to 5. Further improvements of the separation can be achieved through improvements in column efficiency (see Eq. 32). If two or more compounds elute with the same k' value, a change in the mobile phase selectivity is required. Since only three less polar solvents are commonly used in RPLC (methanol, acetonitrile, and tetrahydrofuran), a change in solvent has a relatively limited effect on the selectivity. However, by holding the solvent strength constant while changing the organic modifier (e.g., from methanol to tetrahydrofuran), sufficient changes in selectivity can be achieved. To calculate the volume fraction of the new organic solvent, the following formula can be used (4):

$$\phi_c = \frac{\phi_b(P'_w - P'_b)}{(P'_w - P'_c)} \tag{10}$$

where P'_w is the solvent strength for water, P'_b is the strength of the solvent that gave inadequate selectivity, P'_c is the new solvent, and ϕ_b and ϕ_c are the volume fractions of the two solvents.

By defining another reversed-phase (RP) solvent strength parameter, S, given in Table 2, Eq. 10 can be rewritten in the following form (4):

$$\phi_c = \phi_b \cdot \frac{S_b}{S_c} \tag{11}$$

Table 3   Solvent Strength (RPLC)[a]

Water (0.0)	Dioxane (3.5)
Methanol (3.0)	Ethanol (3.6)
Acetonitrile (3.1)	i-Propanol (4.2)
Acetone (3.4)	Tetrahydrofuran (4.4)

[a]Reproduced from reference 3 with permission.

Use of this equation makes possible calculation of the volume fraction of the desired solvent c, which is necessary to affect the selectivity while maintaining the solvent strength constant.

3 BAND BROADENING

3.1 On-Column Broadening

One of the undesirable aspects of LC in trace analysis is the severe dilution of the sample during its separation. The column acts as a "dilution device," and even with a pluglike injection, the sample will be eluted in a volume larger than its original volume. Thus the dispersion of the band is inevitable, and the elution band will approximate a Gaussian curve. Figure 5 shows a Gaussian-shaped peak and the relationships between peak width and the standard deviation (σ). The dilution factor (DF) arising from the column can be expressed as the ratio of the peak maximum concentration, C_{max}, to the injected concentration (4):

$$ DF = \frac{C_{inj}}{C_{max}} = \frac{\sqrt{\pi/\delta}\, V_{peak}}{V_{inj}} = \frac{\sqrt{2\pi}\, V_{col}\, E_{tot}(1 + k')}{V_{inj}\, N^{1/2}} \quad (12) $$

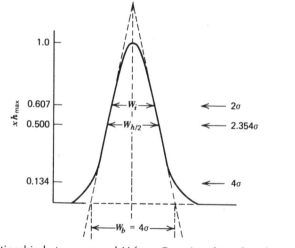

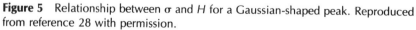

Figure 5   Relationship between σ and H for a Gaussian-shaped peak. Reproduced from reference 28 with permission.

where the peak volume ($V_{peak} = 4\sigma$) is defined as the the base width of the peak elutes, V_{col} is the volume E_{tot} is the total porosity, and N is the number of th column efficiency. Thus if $V_{col} = 2000\ \mu l$, $k' = 2$, the dilution factor calculated from Eq. 12 is 150. T extent of dilution for a typical analytical injection (4). T broadening results from intra- and extracolumn effects.

The full theoretical treatment of the sources of bar beyond the scope of this book and can be found elsewhe ever, it is generally accepted that the overall column b can be expressed by the "height equivalent to a theoreti HETP). This notation was introduced by Martin and Syn based on the analogy between chromatography and distillat tity H is composed of four terms:

$$ H \text{ or } HETP_{tot} = H_{\text{eddy diffusion}} \qquad + H_{\text{longitudinal dif}} $$
$$ H_{\text{resistance to mass transfer}} + H_{\text{extracolumn eff}} $$

The origin of different processes responsible for the overall ening are illustrated schematically in Fig. 6. The individual c can be summarized as follows:

3.1.1 Eddy Diffusion

The term "eddy diffusion" arises from the differences in the mo velocity due to different flow paths that the solute molecules (through the particle bed. Since the solute molecules in the wi will move faster than those in the narrow paths, the initial ba shown in (Fig. 6A), will be dispersed. This type of band broad a function of the tortuosity of the flow pattern through a chromat bed. It should be pointed out, however, that under dynamic co prevailing in the column, solute molecules may also diffuse l (from one streamline to another).

3.1.2 Longitudinal Diffusion (Molecular Diffusion)

Longitudinal diffusion, not shown in Fig. 6, is a band broadening pr that results from diffusion of solute molecules in the direction o the stagnant mobile phase to reach the stationary phase, those which diffuse a longer distance will lag behind the ones which diffuse only a short distance.

3.1.4 Stationary Phase Mass Transfer

The term "stationary phase mass transfer" arises from differences in the residence time of the solute molecules in the stationary phase; molecules that diffuse deeper into the stationary phase will spend more time before they return to the mobile phase and will thus be left behind the main stream of molecules.

The complete van Deemter expression for the plate height (17), developed for GC, assumes that the individual contributions to band broadening are independently additive:

$$H = 2\lambda d_p + \frac{2\lambda D_m}{u} + q\frac{k'd_f^2 u}{(1 + k')^2 D_s} + \frac{Wf(k')d_p^2 u}{D_m} \qquad (14)$$

$$+ \text{ contribution due to stagnant mobile phase}$$

where λ = geometrical factor pertaining to the packing structure
d_p = particle diameter
D_m = diffusion coefficient of the volume in the mobile phase
u = linear velocity
q = geometric factor that takes into account the nature of the column packing
d_f = thickness of the stationary phase
D_s = diffusion coefficient of the solute in the stationary phase
W = constant that takes into account the geometric effect of the packing structure
$f(k')$ = some function of the capacity factor

Generally, the dependence of column efficiency (H and N values) on the experimental parameters can be summarized as follows:

1 H decreases with decreasing particle size of column packing.
2 H decreases with decreasing flow rate.

3 H decreases with decreasing viscosity of the mobile phase and increasing temperature.

4 H is smaller for small solute molecules.

Equation 18 can also be written in the form of

$$H = A + \frac{B}{u} + C_m u + C_s u \tag{15}$$

where C_m is the resistance to mass transfer in the mobile phase, and C_s is the resistance to mass transfer in the stationary phase.

It should be noted that according to Giddings (1), the terms for eddy diffusion (A) and resistance to mass transfer are not independent and therefore may be combined to give:

$$H = \frac{B}{u} + C_s u + \frac{1}{(1/A)(1/C_m u)} \tag{16}$$

Giddings (6) has also suggested that *the theoretical plate height* (HETP or H) should be linked to the particle diameter (d_p). This led to the introduction of *reduced plate height* (h), *reduced fluid velocity* (v), and the *column resistance parameter* (Φ). These parameters, popularized by Knox (9,18–20), offer a simplified method for optimizing chromatographic performance and comparing analyses carried out under different conditions. The reduced parameters are defined by the following equations:

$$\text{Reduced plate height } h = \frac{H}{d_p} \tag{17}$$

$$\text{Reduced fluid velocity } v = \frac{u d_p}{D_m} \tag{18}$$

$$\text{Column resistance parameter (1,9a) } \Phi = \frac{\Delta_p d_p^2}{u \eta L} \tag{19}$$

The parameters in Eq. 23 are as follows: η = solvent viscosity, L = column length, and Δ_p = pressure drop across the column. All other symbols are the same as in the previous expressions. Low values of L and η are necessary for a column to operate efficiently.

the stagnant mobile phase to reach the stationary phase, those which diffuse a longer distance will lag behind the ones which diffuse only a short distance.

3.1.4 Stationary Phase Mass Transfer

The term "stationary phase mass transfer" arises from differences in the residence time of the solute molecules in the stationary phase; molecules that diffuse deeper into the stationary phase will spend more time before they return to the mobile phase and will thus be left behind the main stream of molecules.

The complete van Deemter expression for the plate height (17), developed for GC, assumes that the individual contributions to band broadening are independently additive:

$$H = 2\lambda d_p + \frac{2\lambda D_m}{u} + q \frac{k' d_f^2 u}{(1 + k')^2 D_s} + \frac{W f(k') d_p^2 u}{D_m}$$

$$+ \text{ contribution due to stagnant mobile phase}$$

(14)

where λ = geometrical factor pertaining to the packing structure

d_p = particle diameter

D_m = diffusion coefficient of the volume in the mobile phase

u = linear velocity

q = geometric factor that takes into account the nature of the column packing

d_f = thickness of the stationary phase

D_s = diffusion coefficient of the solute in the stationary phase

W = constant that takes into account the geometric effect of the packing structure

$f(k')$ = some function of the capacity factor

Generally, the dependence of column efficiency (H and N values) on the experimental parameters can be summarized as follows:

1 H decreases with decreasing particle size of column packing.
2 H decreases with decreasing flow rate.

3 H decreases with decreasing viscosity of the mobile phase and increasing temperature.

4 H is smaller for small solute molecules.

Equation 18 can also be written in the form of

$$H = A + \frac{B}{u} + C_m u + C_s u \qquad (15)$$

where C_m is the resistance to mass transfer in the mobile phase, and C_s is the resistance to mass transfer in the stationary phase.

It should be noted that according to Giddings (1), the terms for eddy diffusion (A) and resistance to mass transfer are not independent and therefore may be combined to give:

$$H = \frac{B}{u} + C_s u + \frac{1}{(1/A)(1/C_m u)} \qquad (16)$$

Giddings (6) has also suggested that *the theoretical plate height* (HETP or H) should be linked to the particle diameter (d_p). This led to the introduction of *reduced plate height* (h), *reduced fluid velocity* (v), and the *column resistance parameter* (Φ). These parameters, popularized by Knox (9,18–20), offer a simplified method for optimizing chromatographic performance and comparing analyses carried out under different conditions. The reduced parameters are defined by the following equations:

$$\text{Reduced plate height } h = \frac{H}{d_p} \qquad (17)$$

$$\text{Reduced fluid velocity } v = \frac{u d_p}{D_m} \qquad (18)$$

$$\text{Column resistance parameter (1,9a) } \Phi = \frac{\Delta_p d_p^2}{u \eta \, L} \qquad (19)$$

The parameters in Eq. 23 are as follows: η = solvent viscosity, L = column length, and Δ_p = pressure drop across the column. All other symbols are the same as in the previous expressions. Low values of L and η are necessary for a column to operate efficiently.

(P') values of a mixture of two pure solvents can be calculated using the following formula (4):

$$P' = \phi_1 P_1' + \phi_2 P_2' \tag{9}$$

where ϕ_1 and ϕ_2 are the volume fractions of each solvent in a mixture, and P_1' and P_2' are the polarity indices of the pure solvents.

By selecting an appropriate solvent pair for a particular separation, the k' values for the sample component will be in the range of 2 to 5. Further improvements of the separation can be achieved through improvements in column efficiency (see Eq. 32). If two or more compounds elute with the same k' value, a change in the mobile phase selectivity is required. Since only three less polar solvents are commonly used in RPLC (methanol, acetonitrile, and tetrahydrofuran), a change in solvent has a relatively limited effect on the selectivity. However, by holding the solvent strength constant while changing the organic modifier (e.g., from methanol to tetrahydrofuran), sufficient changes in selectivity can be achieved. To calculate the volume fraction of the new organic solvent, the following formula can be used (4):

$$\phi_c = \frac{\phi_b(P_w' - P_b')}{(P_w' - P_c')} \tag{10}$$

where P_w' is the solvent strength for water, P_b' is the strength of the solvent that gave inadequate selectivity, P_c' is the new solvent, and ϕ_b and ϕ_c are the volume fractions of the two solvents.

By defining another reversed-phase (RP) solvent strength parameter, S, given in Table 2, Eq. 10 can be rewritten in the following form (4):

$$\phi_c = \phi_b \cdot \frac{S_b}{S_c} \tag{11}$$

Table 3   Solvent Strength (RPLC)^a

Water (0.0)	Dioxane (3.5)
Methanol (3.0)	Ethanol (3.6)
Acetonitrile (3.1)	i-Propanol (4.2)
Acetone (3.4)	Tetrahydrofuran (4.4)

^{a}Reproduced from reference 3 with permission.

Use of this equation makes possible calculation of the volume fraction of the desired solvent c, which is necessary to affect the selectivity while maintaining the solvent strength constant.

3 BAND BROADENING

3.1 On-Column Broadening

One of the undesirable aspects of LC in trace analysis is the severe dilution of the sample during its separation. The column acts as a "dilution device," and even with a pluglike injection, the sample will be eluted in a volume larger than its original volume. Thus the dispersion of the band is inevitable, and the elution band will approximate a Gaussian curve. Figure 5 shows a Gaussian-shaped peak and the relationships between peak width and the standard deviation (σ). The dilution factor (DF) arising from the column can be expressed as the ratio of the peak maximum concentration, C_{max}, to the injected concentration (4):

$$\text{DF} = \frac{C_{inj}}{C_{max}} = \frac{\sqrt{\pi/\delta}\ V_{peak}}{V_{inj}} = \frac{\sqrt{2\pi}\ V_{col}\ E_{tot}(1 + k')}{V_{inj}\ N^{1/2}} \tag{12}$$

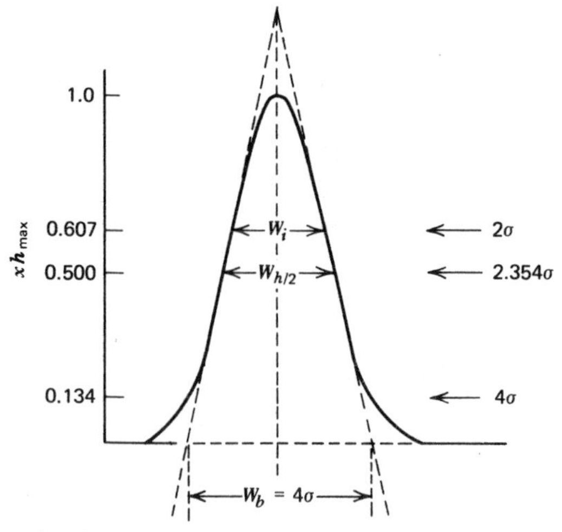

Figure 5 Relationship between σ and H for a Gaussian-shaped peak. Reproduced from reference 28 with permission.

where the peak volume ($V_{peak} = 4\sigma$) is defined as the volume over which the base width of the peak elutes, V_{col} is the volume of empty column, E_{tot} is the total porosity, and N is the number of theoretical plates or column efficiency. Thus if $V_{col} = 2000$ μl, $k' = 2$, and $N = 10,000$, the dilution factor calculated from Eq. 12 is 150. This illustrates the extent of dilution for a typical analytical injection (4). The observed band broadening results from intra- and extracolumn effects.

The full theoretical treatment of the sources of band broadening is beyond the scope of this book and can be found elsewhere (6–17). However, it is generally accepted that the overall column band broadening can be expressed by the "height equivalent to a theoretical plate" (H or HETP). This notation was introduced by Martin and Synge (12a) and is based on the analogy between chromatography and distillation. The quantity H is composed of four terms:

$$H \text{ or } HETP_{tot} = H_{\text{eddy diffusion}} + H_{\text{longitudinal diffusion}} +$$
$$H_{\text{resistance to mass transfer}} + H_{\text{extracolumn effects}} \tag{13}$$

The origin of different processes responsible for the overall band broadening are illustrated schematically in Fig. 6. The individual contributions can be summarized as follows:

3.1.1 Eddy Diffusion

The term "eddy diffusion" arises from the differences in the mobile phase velocity due to different flow paths that the solute molecules (X) follow through the particle bed. Since the solute molecules in the wider paths will move faster than those in the narrow paths, the initial bandwidth, shown in (Fig. 6A), will be dispersed. This type of band broadening is a function of the tortuosity of the flow pattern through a chromatographic bed. It should be pointed out, however, that under dynamic conditions prevailing in the column, solute molecules may also diffuse laterally (from one streamline to another).

3.1.2 Longitudinal Diffusion (Molecular Diffusion)

Longitudinal diffusion, not shown in Fig. 6, is a band broadening process that results from diffusion of solute molecules in the direction of the

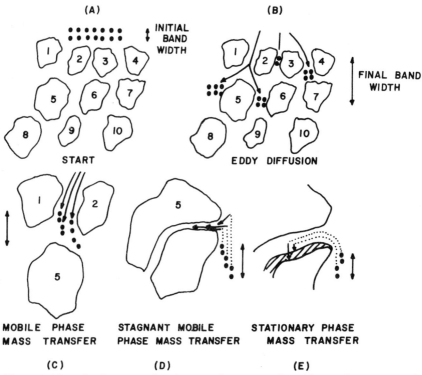

Figure 6 Contributions to molecular spreading. Reproduced from reference 27 with permission.

mobile phase flow. The term becomes significant only under stopped-flow conditions and at low flow rates. Because of the low solute diffusivity in liquids, this term is usually negligible in HPLC.

3.1.3 Mobile Phase Mass Transfer

Mobile phase mass transfer can be subdivided into two terms: the moving mobile phase mass transfer (Fig. 6C) and the stagnant mobile phase mass transfer (Fig. 6D). The moving mobile phase mass transfer results from the differences in the mobile phase velocity within a single streamline: solute molecules in the midstream will move faster than those close to the particle surface. The band broadening due to the stagnant mobile phase mass transfer results from the existence of stagnant mobile phase in the intraparticle void volume. Since solute molecules diffuse through

The reduced plate height, h, for a nonsorbed solute in a column packed with spherical particles can be expressed by the following relationship (8,16):

$$h = \frac{2\gamma}{\nu} + \frac{2}{1 + \omega\nu^{1/3}} + \frac{\kappa k_0^2}{(1 + k_0)^2}\nu^{2/3} + \frac{\Theta k_0}{30(1 + k_0)^2}\nu \quad (20)$$

where γ, η, ω, and κ are structural parameters of the column packing, Θ is the tortuosity factor for porous particles, k_0 is the ratio of accessible intraparticulate void volume to the interstitial void space in the column, and ν is the reduced velocity. The individual terms in Eq. 24 account for the longitudinal diffusion, eddy diffusion, resistance to mass transfer at the particle boundary, and interparticulate resistance to mass transfer.

As proposed by Done et al. (21,22), the dependence of the reduced plate height on reduced velocity can be expressed using a simple semi-empirical formula:

$$h = \frac{B}{\nu} + A\nu^{0.33} + C\nu \quad (21)$$

Usually, with a well-packed column, the constant A will have a value around unity. The term B can be estimated from the obstruction factors (particle shape and porosity) for diffusion in the two phases, the k' value, and the diffusion coefficients, and is typically between 1.5 and 2.0. The C factor, which is a measure of the dispersion in the stationary and stagnant mobile phases, will have a theoretical minimum of 10^{-2} for columns packed with totally porous particles and slightly retained solutes.

The general relationship of reduced plate height versus reduced velocity, known as the Knox plot, is illustrated in Fig. 7.

The values of A, B, and C for a given column can be obtained by curve fitting. For reliable data, the curve should be constructed over at least a 100-fold range of reduced velocities.

Further simplifications of Eq. 21 under specific velocity conditions can be found in the work of Done, Kennedy, and Knox (22).

3.2 Extracolumn Band Broadening

Although much attention is paid to band broadening caused by intracolumn effects, extracolumn effects are often overlooked. Large volumes

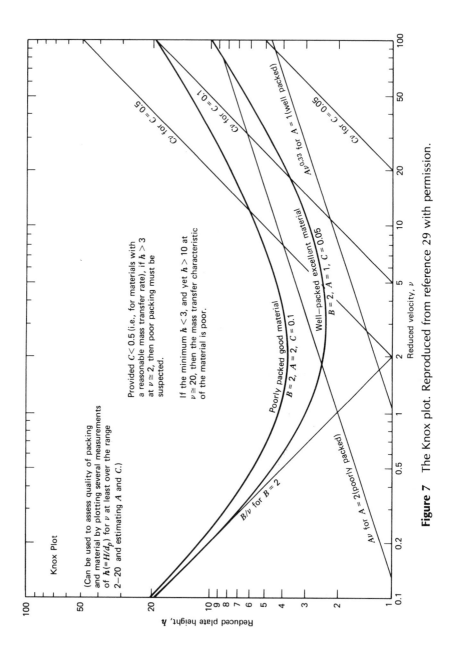

Figure 7 The Knox plot. Reproduced from reference 29 with permission.

in fittings, connecting tubing, and detector cells, as well as the method of sample introduction, can greatly reduce the efficiency of a separation (9,12,23–25). Therefore, it is absolutely essential to use fittings and detector cells that are well swept and have zero dead volume.

Since the chromatographic resolution and accuracy of the analysis can be affected by extracolumn effects, it is important to be aware of the individual contributions to band broadening that arise from separate parts of the chromatographic system. Individual contributions or variances (σ^2) can be calculated according to the following expressions:

$$\sigma_v^2 = F^2\sigma_b^2 \tag{22}$$

or

$$\sigma_t^2 = \frac{t_R^2}{N} \tag{23}$$

The subscripts v and t are used to denote the volume and time units, respectively.

The overall variance ($\sigma_{v_{tot}}^2$) of the elution curve is composed of independent variances arising from the column effects, injector, connecting tubing, and dispersion in the detector. The second moments or variances of individual contributions to the total observed variance are additive (23):

$$\sigma_{v_{tot}}^2 = \sigma_{v_{col}}^2 + \sigma_{v_{inj}}^2 + \sigma_{v_{conn}}^2 + \sigma_{v_{det}}^2 + \sigma_{v_{other}}^2 \tag{24}$$

where the subscript v refers to the variance on the volume basis.

If the extracolumn effects are negligible, the total observed variance is reduced to the following expression:

$$\sigma_{v_{tot}}^2 = \sigma_{v_{col}}^2 \tag{25}$$

However, in most cases, $\sigma_{v_{tot}}^2$ is larger than $\sigma_{v_{col}}^2$, and the chromatographic resolution is lower compared to the inherent column capability.

Knox (9) has reported maximum allowable lengths that can be used with connecting tubing of different diameters. In general, the length of connecting tubing (ℓ in cm) that can be used with a maximum bandwidth increase of 5% can be calculated from the following expression (24):

$$\ell = \frac{40V_R^2 D_m}{11F_v d^4 N} \qquad (26)$$

where V_R is the solute retention volume (ml), D_m is the solute diffusion coefficient (cm^2/sec), F is the flow rate (ml/sec), d is the diameter of the tubing (cm), and N is the column plate number. The existence of extra-column effects is usually manifested in band tailing and lower plate counts of the early eluting peaks ($k' < 2$) than of those which elute later ($k' > 2$).

High-performance work also requires careful design of the injection device. The proper injection technique and placing the sample as close as possible to the head of the column (without disturbing the packing) will reduce band dispersion. Extracolumn broadening can also arise from large volumes and/or concentrations of the injected sample. This results in increasing H values with a concomitant decrease in efficiency. With the highly efficient microparticulate columns in use today, this effect is more evident from the efficiency values than the retention times. There-fore, it is important to establish the dependence of both efficiency and retention time on the sample size. However, if the levels of compounds of interest in a given sample are low, it is possible to inject larger amounts of sample without loss of efficiency. This is applicable in cases where the total volume injected is less than approximately one-third of the first peak of interest in the chromatogram (25). Individual contributions to the overall extracolumn band broadening will be discussed in Section III.

4 COLUMN EFFICIENCY

The prime objective of a successful RPLC analysis is to obtain efficient separations and narrow bands. A measure of column efficiency is the number of theoretical plates of the column (N), which is given by the following expression:

$$N = 16 \left(\frac{t_R}{W_b} \right)^2 \qquad (27)$$

or

$$N_{\text{eff}} = 16 \left(\frac{t_R - t_0}{W_b} \right)^2 = N \left(\frac{k'}{1 + k'} \right)^2 \qquad (28)$$

where W_b is the base width of the peak. By inspection of Eq. 27, it is obvious that if N remains constant, the bandwidth (W_b) will increase with increasing retention time (t_R). Thus, with isocratic elution, late eluting peaks will have significantly reduced peak heights, which makes the detection of trace amounts difficult. However, this can be prevented by means of gradient elution, since peaks in the chromatogram can then be of equal width. An alternative expression for calculating column efficiency avoids errors associated with the drawing of tangents for the determination of W_b:

$$N = 5.54 \left(\frac{t_R}{W_{1/2}} \right)^2 \tag{29}$$

where $W_{1/2}$ is the width of the chromatographic band at half-height. To account for the influence of the column length (L) on the number of theoretical plates, the height equivalent to a theoretical plate (H), rather than N, is usually reported:

$$H(\text{mm}) = \frac{L(\text{mm})}{N} \tag{30}$$

It should be noted that the lower the plate height, the higher the N value and the better the column efficiency.

5   RESOLUTION AND CONTROL OF SEPARATION

In order for two peaks from solutes of different chemical types to be separated, their bands must be moved sufficiently apart during their passage through the column. In addition, the widths must remain narrow if compounds are to be eluted as discrete peaks. The resolving power of the column, which relates the width of the eluted peaks (W) to the distance between their maxima, is referred to as the *resolution*, R_s, and is given by the following equation:

$$R_s = 2 \left(\frac{t_{R_2} - t_{R_1}}{W_1 + W_2} \right) \tag{31}$$

The subscript 2 refers to the peak with longer retention times.

If the bands have Gaussian shape, the separation of the maxima of the two peaks will increase in proportion to the distance migrated, while the bandwidth will increase as the square root of the distance migrated. When $R_s = 1$, the overlap between the two bands is 4σ. If $R_s = 1.5$, the separation is essentially complete (overlap less than 1%, 6σ separation). It should be pointed out that if one of the solutes is present only at a trace level, the overlap of the larger peak will be significant, and it will be difficult to distinguish the bands, even when the resolution factor is larger than 1. The change in separation as a function of R_s and the peak height ratio is illustrated in Fig. 8 (26).

Although Eq. 31 is useful for estimating the resolving power of a chromatographic system and defining the separation, it does not show how the resolution can be improved and controlled. By integrating individual expressions such as column efficiency (N), selectivity (α), and capacity factor (k') into the resolution expression for chromatographic

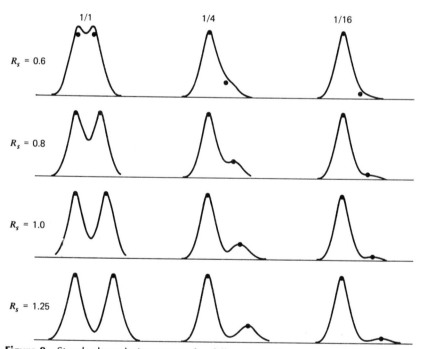

Figure 8   Standard resolution curves for different relative band concentrations. Reproduced from reference 27 with permission.

performance, a general, more useful resolution expression can be obtained:

$$R_s = \frac{1}{4} \left(\sqrt{N} \right) \left(\frac{\alpha - 1}{\alpha} \right) \left(\frac{k_2'}{k_2' + 1} \right) \qquad (32)$$

or

$$R_s = \frac{1}{4} \sqrt{N}_{\text{eff}} \left(\frac{\alpha - 1}{\alpha} \right) \qquad (33)$$

Both equations take into account the kinetic factor (N) and the thermodynamic factors (α and k'), which can be varied to improve LC separations. It should be pointed out that Eqs. 32 and 33 are valid only for Gaussian peaks (linear partition isotherm). Equation 32 can also be used to calculate the optimum column length (N and H) that is necessary for the desired resolution. The control of resolution by means of different parameters shown in Eq. 32 is illustrated in Fig. 9. A decrease in the k' values of the original bands will cause a decrease in resolution. Conversely, increased k' values will lead to an increase in resolution and a concomitant increase in analysis time and a decrease in peak heights. Improved efficiency (larger plate number, N) results in improved separation of the two bands and narrower peak widths. If N is increased by decreasing the solvent flow rate or by increasing the column length, the analysis time will also be increased. Finally, if the selectivity is increased, the resolution will be improved without significant changes in the analysis time or peak heights. Each factor will now be discussed separately.

In order to improve resolution by changing the efficiency of the column (N), column length, particle size, or flow rate may be changed. The most practical way to increase the efficiency and resolution is by decreasing the flow rate or column pressure. This will, however, be accompanied by a concomitant increase in separation time. It is also possible to increase N by increasing the column length while keeping the column pressure constant (by decreasing flow). In practice, this is achieved by connecting several columns in a series, and the separation times are shorter than with fixed column lengths and decreased column pressures. Resolution can also be improved by simultaneous increase in column length and pressure, without increasing the analysis time. α and k' factors, which are directly related to the solute-solvent-solvent interactions, can also be used to affect

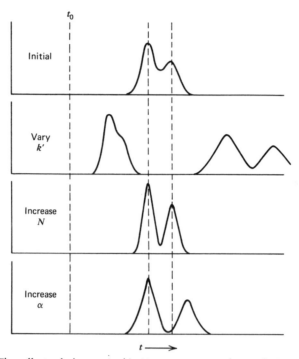

Figure 9   The effect of change in k', N, or α on sample resolution. Reproduced from reference 27 with permission.

the resolution. The change in the thermodynamics of the system can be brought about by varying the phase ratio (V_s/V_m), by using different phase systems, by changing the solvent composition and pH, and by means of secondary equilibria. Improvements of R_s values by increasing α, while maintaining optimal k' values, are usually achieved by trial and error. However, general guidelines are available. The most effective option is the change in the nature of the mobile phase (use of different solvents), which enables improvements in selectivity with no detrimental effect on the k' values. Separations of samples that contain ionizable solutes can be optimized by changing the pH of the mobile phase. If an improvement in α values is accompanied by deterioration of the k' values, changes in solvent strength (variations in the proportions of the mobile phase components) usually give satisfactory results. Changing the stationary phase as a way of improving the α values is inconvenient and is rarely used.

While an increase in temperature has little effect on α values in LLC and LSC, its use is more promising in ion-exchange and ion-pair chromatography. In the two latter methods, increased temperatures usually lead to a decrease in k' values, in addition to improvements in the α values. Therefore, it is often necessary to decrease the solvent strength in order to counteract the effect of temperature on the k' values. Finally, drastic changes in α values can be achieved by means of secondary equilibria: complexation with silver(I) ions, borate-*vic*-diol reactions, bisulfite-aldehyde reactions, and so on have been successfully used for enhancement of selectivity. By controlling the last term in Eq. 32 [$k'/(k' + 1)$], which denotes the fraction of solute molecules in the stationary phase, resolution can be greatly affected.

Changes in the k' values are brought about by variations in the solvent strength. Polar solvents such as water give rise to low k' values, whereas less polar solvents such as methanol and acetonitrile result in larger k' values. Resolution of mixtures containing compounds of widely different k' values can be optimized by means of gradient elution, where the composition of the mobile phase is continuously varied during separation, in order to achieve an optimal range of $1 < k' < 10$.

6 ANALYSIS TIME

A factor of considerable interest to the chromatographer is the time required to perform a given separation. The main objective is to achieve the best separation in minimum time.

It can be shown that the time of analysis (elution of the last peak) can be defined by the following expression:

$$t_R = 16R_s^2\left(\frac{\alpha}{\alpha - 1}\right)\left(\frac{(1 + k')^3}{k'^2}\right)\left(\frac{H}{u}\right) \tag{34}$$

From Eq. 34 it is evident that if all other variables are held constant, doubling the resolution causes a fourfold increase in analysis time. Since the optimal k' values lie in the range of 2 to 5, it is not always possible to achieve an even degree of separation of all the components in a mixture if a mobile phase of constant composition (isocratic elution) is used. To eliminate a large spread of k' values, k' programming is used. This usually involves gradual changes in eluent strength (gradient elution). Less often, flow programming, temperature programming, and coupled columns are used for this purpose.

REFERENCES

1. M. J. M. Wells and C. R. Clark, *Anal. Chem.*, **53**, 1341 (1981).
1a. L. R. Snyder, in *Techniques of Chemistry*, 2nd ed., Vol. 3, Part 1, A. Weissberger and E. S. Perry, Eds., Wiley-Interscience, New York, 1979, Chap. 2.
2. L. Rohrschneider, *Anal. Chem.*, **45**, 1241 (1973).
3. L. R. Snyder, *J. Chromatogr.*, **92**, 223 (1974); *J. Chromatogr. Sci.*, **16**, 223 (1978).
4. L. R. Snyder and J. J. Kirkland, *Introduction to Modern Liquid Chromatography*, 2nd ed., Wiley-Interscience, New York, 1979, Chap. 6.
5. L. R. Snyder, *Principles of Adsorption Chromatography*, Dekker, New York, 1976, Chap. 8.
6. J. C. Giddings, *Dynamics of Chromatography, Part I, Principles and Theory*, Dekker, New York, 1965.
7. E. Grushka, R. Snyder, and J. H. Knox, *J. Chromatogr. Sci.*, **13**, 25 (1975).
8. C. Horváth and W. Melander, *J. Chromatogr. Sci.*, **15**, 393 (1977).
9. J. H. Knox, *J. Chromatogr. Sci.*, **15**, 352 (1977).
10. R. P. W. Scott, *Contemporary Liquid Chromatography*, Wiley-Interscience, New York, 1976, p. 47.
11. J. C. Sternberg, *Adv. Chromatogr.*, **2**, 205 (1966).
12. J. H. Knox and A. Pryde, *J. Chromatogr.*, **112**, 171 (1975).
12a. A. J. P. Martin and R. L. M. Synge, *Biochem. J.*, **35**, 1358 (1941).
13. G. J. Kennedy and J. H. Knox, *J. Chromatogr. Sci.*, **10**, 549 (1972).
14. J. F. Huber, *Ber. Bunsenges Phys. Chem.*, **77**, 179 (1973).
15. I. Halász, H. Schmidt, and P. Vogtel, *J. Chromatogr.*, **126**, 19 (1976).
16. C. Horváth and H.-J. Lin, *J. Chromatogr.*, **126**, 401 (1976).
17. J. J. Van Deemter, F. J. Zuiderweg, and A. Klinkenberg, *Chem. Eng. Sci.*, **5**, 271 (1956).
18. P. A. Bristow and J. H. Knox, *Chromatographia*, **10**, 279 (1977).
19. J. H. Knox and J. F. Parcher, *Anal. Chem.*, **41**, 1599 (1969).
20. G. L. Laird, J. Jurand, and J. H. Knox, *Proc. Soc. Anal. Chem.*, **11**, 310 (1974).
21. J. N. Done and J. H. Knox, *J. Chromatogr. Sci.*, **10**, 606 (1972).
22. J. N. Done, G. J. Kennedy, and J. H. Knox, in *Gas Chromatography*, 1972, S. G. Perry and E. R. Adlard, Eds., Applied Science Publ., Barking, Essex, 1973, pp. 145–155.
23. J. C. Sternberg, in *Advances in Chromatography*, Vol. 2, J. C. Giddings et al., Eds., Dekker, New York, 1966, p. 206.
24. R. P. W. Scott and P. Kucera, *J. Chromatogr. Sci.*, **9**, 641 (1971).
25. J. N. Done, *J. Chromatogr.*, **125**, 43 (1976).
26. L. R. Snyder, *J. Chromatogr. Sci.*, **10**, 200 (1972).

27. L. R. Snyder and J. J. Kirkland, *Introduction to Modern Liquid Chromatography,* 2nd ed., Wiley-Interscience, New York, 1979, Chaps. 2,6.

28. C. F. Simpson, *Practical High Performance Liquid Chromatography,* Heyden, London, 1976, p. 9.

29. P. A. Bristow, *LC in Practice,* HETP Publ., 10 Langley Drive, Handforth, Wilmslow, Cheshire, U.K., 1976.

III Instrumentation

During the past decade, significant advances have been made in the HPLC instrumentation. The operational systems presently in use are of considerable complexity and sophistication. Because of continuous developments, detailed layouts of chromatographic systems are rapidly outdated. Since most improvements are only refinements of the essential components that make up an HPLC system, only the functional components will be described in general terms.

Because the number of manufacturers of complete HPLC systems or components is growing constantly, selection of an HPLC system requires a judicious choice. Since most components are still rather costly, extra expenses can be avoided if specialized equipment is not necessary. The choice of an instrument will be dictated by the following factors:

1 Complexity of sample matrix (isocratic elution vs. gradient).
2 Sample size (analytical vs. preparative).
3 Sensitivity of detection.
4 Number of analysis to be performed.
5 Frequency of use.

Currently, it is advisable to purchase individual components in order to achieve greater flexibility and to be able to update the instrument when new components are available. The basic components common to all liquid chromatographs are outlined in Table 4. These components are schematically illustrated in Fig. 10. Since all components contribute to the overall performance of the instrument, each one will be discussed separately.

In addition to the major components listed in Table 4, tubing, tube fitting, and the material out of which the instrument is constructed are also important since they influence the performance and/or instrument lifetime. Most commercially available chromatographs are made out of stainless steel, grade ANSI 316, which is highly resistant to corrosion by oxidizing agents, acids, bases, and organic solvents. The presence of

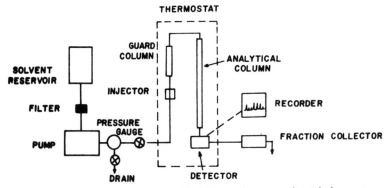

Figure 10 Schematic representation of a high-performance liquid chromatograph.

a protective oxide layer resists oxidizing agents and stops further corrosion. Mineral acids, halide ions, and anions of some carboxylic acids can disrupt this layer, causing some deterioration of the inner metal surface. Therefore, these compounds should be avoided if possible and the equipment should be rinsed with water after use, if it is essential to use these compounds to obtain a given separation. Other materials used for the fabrication of chromatographic equipment are polytetrafluoroethylene (PTFE), glass, and certain types of plastics. However, some types of plastics cannot be used with specific solvents.

1 CONNECTING TUBING

Connecting tubing is also constructed of stainless steel. Because of its role in extracolumn broadening, its dimensions are precisely controlled.

Table 4 Components of a Liquid Chromatograph

1	Solvent supply system	Solvent reservoirs, solvent degassing system, pump, gradient elution programmer
2	Sample introduction system	Septum-type syringe injector, septumless sryinge injector, sample valve, auto sampler
3	Precolumn and analytical column	
4	Temperature control	
5	Detectors	
6	Fraction collector	Manual or automatic
7	Data handling devices	Recorder, integrator, computer

It should be pointed out that the diameter of the tubing up to the sample injector is not critical and its dimensions are usually 0.75 mm (0.030 in.) I.D. Much more rigorous demands are placed on connecting tubing and fittings beyond the injection point. The fittings should have a zero dead volume, and the tubing should have a maximum inner diameter of 0.25 mm (0.010 in.). In preparative work, the dimensions of the tubing are larger, due to the need for higher flow rates.

1.1   Band Broadening due to Connecting Tubing

The degree of dispersion resulting from tubing and connectors has been treated comprehensively by Scott and Kucera (5) and Sternberg (6). The overall dispersion volume (v_i) due to individual connectors can be expressed by the following equation:

$$v_i^2 = v_1^2 + v_2^2 + \cdots \tag{35}$$

In order to evaluate the dispersion caused by connecting tubing, two extreme cases can be considered: fully flushed connecting tubing and exponentially flushed connecting tubing. However, that real systems will be a combination of the two.

1.1.1   Fully Flushed Connecting Tubing

The contribution to the peak volume from axial dispersion in an open tube can be described by the following equation (1):

$$v_{\text{tube}}^2 \approx \frac{0.13Ld^4F_v}{D_m} \tag{36}$$

or

$$\sigma_v^2 = \frac{\pi d^4 L F_v}{384 D_m} \tag{37}$$

where L is the length of tubing, d is the diameter, F is the flow rate, and D_M is the solute diffusion coefficient in the mobile phase. By using typical values of $F_v = 20$ μl/sec and $D_M = 10^{-3}$ mm^2/sec and setting $V_{\text{tube}} < 30$ μl, the equation becomes:

$$d^4L \leqslant 0.34 \text{ mm}^5 \tag{38}$$

Knox (3) lists the maximal lengths and bore sizes of connecting tubing under conditions mentioned in the preceding paragraph.

1.1.2 Exponentially Flushed Connecting Tubing

Laminar flow of solvent along a smooth bore tube is considerably less efficient than completely mixed flushing. The volume variance resulting from flushing a completely mixed volume, Q, can be expressed by the following equation (3):

$$\sigma_v^2 = Q^2 \tag{39}$$

and the corresponding dispersion volume is:

$$v_{\text{conn}} = 4Q \tag{40}$$

Since there may be several connectors, few of which will be fully mixed, the resulting dispersion will be larger than predicted by Eq. 44 (3).

Usually, if the connecting tubing with 0.25-mm bore is shorter than 100 mm, and if zero dead volume connectors (tubing ends butted together) are used, the detector cell will be the main source of extracolumn band broadening.

2 SOLVENT DELIVERY SYSTEM

With microparticulate packings, there is a high-pressure drop across the chromatographic column. Therefore, great demands are placed on the solvent delivery system. Regardless of the type of pump used, certain features are required for satisfactory performance of the liquid chromatograph:

1 Constant flow delivery (0.1–10 ml/min, 2% variation).
2 High maximal pressure.
3 Delivery of discretely variable flow.
4 Minimal pressure fluctuations.
5 Low noise level.
6 Simplicity of operation.
7 Chemical inertness to most commonly used solvents.

The need for flow constancy stems from the fact that most commonly used HPLC detectors are concentration sensitive. Thus the area of the peak, which is used in quantitation, is related to the flow rate of the mobile phase through the following expression:

$$A = \frac{Rm}{F} \qquad (41)$$

A is the peak area, R is the solute response factor, m is the mass of solute, and F is the flow rate. Significant errors may result where peak areas used for quantitation are obtained under varying solvent flow conditions. In scouting for the optimum solvent flow rate, the ease with which the flow rate can be changed is important. In addition, the speed of analysis should not be limited by the maximum operating pressure of the pump.

2.1 Solvent Degassing System

Due to the pressure drop across the column, the solubility of gases dissolved in the mobile phase decreases, leading to the evolution of bubbles. This is particularly obvious with alcohol-water mixtures. The problem becomes more pronounced with increasing inlet pressures. The presence of bubbles in the chromatographic system is undesirable since they cause variations in the flow, lead to degradation of resolution, and cause instability of detectors employing flow-through cells. Since some pumps are equipped with pressure transducers which provide a direct pressure readout, the presence of bubbles in the pump can be detected by oscillations of the pressure indicator. The presence of bubbles within the detector cell is manifested by "spikes" in the chromatogram.

2.2 Pumps

The purpose of the pump is to supply the mobile phase at a precisely controlled flow rate or pressure. Some commercial pumps are equipped with feedback systems that maintain either one of the two parameters at a constant, predetermined level. Pumps are normally divided into two categories: *constant-pressure* and *constant-flow pumps*. The latter type is more common. Depending on the principle of operation, commonly available pumps can also be categorized as follows:

1   Pneumatic pumps.
2   Syringe-type pumps.

3 Reciprocating pumps.

4 Hydraulic amplifier pumps.

2.2.1 Pneumatic Pumps (Constant-Pressure Pumps)

Pneumatic pumps employ the pressure of a gas (He or N_2) either to drive or regulate the pressure of the eluent. They can be subdivided into: (1) direct gas displacement, (2) pneumatic amplifier pumps, and (3) pressure-regulated pumps. These pumps supply the mobile phase at a constant pressure, which in the isocratic operation is equivalent to constant flow. However, in gradient elution, where the viscosity of the mobile phase is continuously changing, or in cases where the permeability of the column changes (due to compression of the bed with time), drifts in retention times may occur. In addition, since most commonly used LC detectors are concentration-sensitive, peak areas can be significantly affected.

1 Direct gas displacement pumps operate by applying the pressure of the gas directly to the surface of the mobile phase or to a suitable collapsible or regular container. Therefore, they are limited by the pressure of the gas cylinder ($\sim$2500 psi). Pumping systems in which the gas is in direct contact with the eluent should have a minimal gas-liquid interface in order to decrease the rate of gas dissolution, which leads to bubbling in the low-pressure part of the chromatographic system (i.e., the detector cell). This can be eliminated by use of collapsible plastic liquid containers, plastic bellows, metal bellows, or fluids such as mercury or high-viscosity oils. A schematic diagram showing some types of gas driven pumps is shown in Fig. 11. The use of these pumps is limited to moderate pressures and the limited volumes of eluent that can be delivered before the reservoir has to be refilled. In addition, gradient elution cannot be used. However, it should be pointed out that these pumping systems are inexpensive and the flow is usually pulseless, which is important for achieving low detection limits.

2 Pneumatic amplifier pumps use the Pascal principle to amplify the pressure: the area of the piston in direct contact with the gas (A_G) is larger than the area of the eluent piston (A_L) and the resulting amplification of the pressure is given by the following expression:

$$P_L = \frac{A_G}{A_L}P_G \tag{42}$$

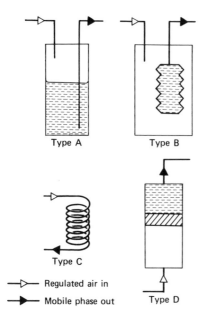

Figure 11 Design of simple pumps using gas pressure as the driving force. In type B, use is made of a collapsible plastic bottle or bellows. In type D, a sliding piston is used. Reproduced from reference 22 with permission.

where P_L is the pressure exerted on the mobile phase, and P_G is the pressure on the gas piston. Thus pressures up to 500 atm can be achieved. Figure 12 shows a typical diagram of this type of pump. A valve system (controlled automatically or manually) permits rapid refilling of the eluent chamber (a few seconds), and separations are not significantly perturbed because of the small solute diffusivity in liquids. However, since the column is momentarily depressurized during refilling, a spurious signal may appear in the chromatogram, the magnitude of which is determined by the pressure and the sensitivity and type of the detector. Although the action of these pumping systems is discontinuous, they still deliver a pulseless flow (except during solvent refilling), and they make possible rapid change of solvents. In addition, flow rates up to 100 ml/min can be delivered under high pressures, which is advantageous in preparative work. The main disadvantage of this type of pump is that it cannot be used in recycling.

3 The pressure-regulated pumps employ a reciprocating piston pump and a pressure regulator, gas pressure regulator, spring pressure regulator, or the continuous gas displacement pump. With gas pressure regulators, a wider range of pressures can be achieved. Con-

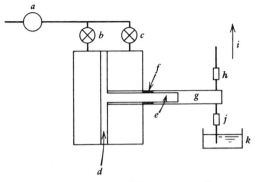

Figure 12 Diagram of a pneumatic amplifier pump. a, Gas pressure controller; b and c, valves; d, gas piston; e, liquid piston; f, piston seal; g, solvent chamber; h, column check valve; i, to column; j, reservoir check valve; k, reservoir. Reproduced from reference 23 with permission.

tinuous gas displacement pumps are usually equipped with a controlled leak section which dampens the flow pulsations.

2.2.2   Syringe-Type Pumps

Syringe-type pumps operate on the principle of solvent displacement from a cylinder (250–500 ml), using a tightly fitted piston that advances at a constant rate. The delivery of the liquid is controlled by electromechanical means. The solvent reservoir must be refilled periodically, but due to its relatively larger volume, several analytical separations can be performed without refilling. The refilling process is longer than with pneumatic pumps since the piston is driven mechanically. A schematic of a syringe-type pump is shown in Fig. 13. It is generally accepted that these pumps deliver a very constant flow in a pulseless operation. At higher operating pressures, this assumption does not hold true since the compressibility

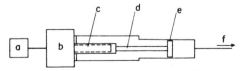

Figure 13 Syringe-type pumps: a, Stepping motor; b, gear box or sprockets roller speed reducer assembly; c, screw; d, plunger piston; e, piston seal; f, to column. Reproduced from reference 16 with permission.

of liquids becomes appreciable. This is particularly true in gradient elution operation, where solvents of different viscosities are often employed.

2.2.3 Reciprocating Pumps

Basically, there are two types of reciprocating pumps: piston reciprocating pumps and diaphragm or membrane reciprocating pumps. In the first type, the piston is in direct contact with the liquid being pumped, whereas in the latter, the movement of the piston is transmitted to a diaphragm or a membrane via a hydraulic system. The pump head is equipped with check valves, synchronized with the piston or the diaphragm, which allow suction and delivery of solvent during alternate strokes. With sinusoidal-drive reciprocating pumps, the resulting flow consists of a series of rapid pulses. To reduce the amplitude of the pulsations in the flow, several possibilities are available: the use of damping devices, by connecting two reciprocating pumps (which operate out of phase) in parallel, and the use of twin piston pumps which employ a constant piston displacement but vary the pumping frequency depending on the flow requirements. Alternatively, some manufacturers offer pumps with two pump heads mounted 180° out of phase (one head is filling while the other is delivering the liquid), which results in considerable smoothing of the flow.

The profiles of the output from a reciprocating pump are shown in Fig. 14. The residual fluctuation can be eliminated by means of high-sensitivity noise filters and damping devices.

The main advantage of reciprocating pumps is their ability to deliver a continuous flow from an external reservoir of unlimited capacity. In addition, they make possible the use of recycle chromatography and the rapid exchange of solvents owing to the small volume of the pump heads.

With the sinusoidal-drive reciprocating pumps, pulse damping devices must be used if a pulseless flow is to be obtained. Conversely, dual-head pumps deliver an almost constant and pulseless flow at high back pressures; however, small cyclic pulsations result from the compressibility of liquids. Some commercial pumps employ micrometric screws to adjust the piston course, which results in constant and pulseless flow.

2.2.4 Hydraulic Amplifier Pumps

These pumps are similar to the pneumatic amplifier pumps, except that they use a low-pressure fluid (oil) instead of a gas to drive the liquid.

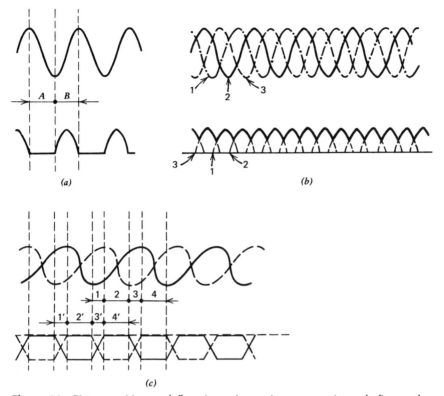

Figure 14 Piston position and flow in reciprocating pumps. In each figure, the upper curve gives the position of the piston and the lower curve gives the flow-rate profile as a function of time. *(a)* Single sinusoidal piston movement head; A, backward stroke; B, forward stroke. *(b)* Three sinusoidal piston movement heads. In the lower diagram the thick solid line indicates the total flow rate. *(c)* Dual-head, special-piston movement pump. 1, Piston A accelerates; 2, piston A moves at constant speed; 3, piston A decelerates; 4, piston A retracts; 1′ to 4′, similar movements of piston B. The broken line indicates the total flow rate. Reproduced from reference 23 with permission.

The pumping system, shown in Fig. 15, consists of low- and high-pressure circuits, interfaced by the intensifiers (180° out of phase), which contain the dual-piston assembly. The delivery of oil to the intensifier and the evacuation to the oil reservoir are controlled by solenoid valves. During the evacuation of oil, the piston in one intensifier is moving backward to refill the solvent chamber, and during intake of oil, the solvent is being delivered.

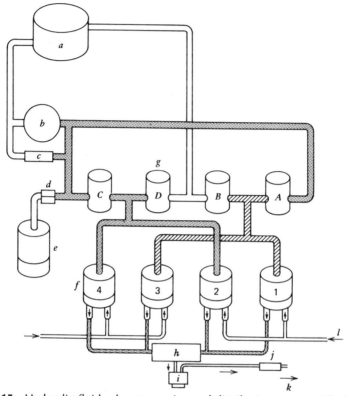

Figure 15 Hydraulic fluid-solvent pumping and distribution systems. Black tubes, pressurized liquid; hatched tubes, alternate. a, Oil reservoir; b, gear pump; c, pressure limiter; d, bleed check; e, low pressure accumulator; f, intensifiers (1, 2, 3, 4); g, solenoid valves (A, B, C, D); h, blending valve; i, pressure solvent. Reproduced from reference 23 with permission.

The functioning of solenoid valves causes variations in the flow rate. In addition, recycle chromatography is not possible with these pumping systems.

2.2.5 Feedback Systems

For a successful chromatographic separation, pumps are required to operate at constant hydrodynamic conditions. In order to ensure the constancy of flow and/or pressure, special electronic feedback systems are incorporated into the pumping system. By measuring an operational parameter such as the column-inlet pressure, the flow rate, or the rotation speed of the motor and comparing it with the desired value, the difference can be amplified and used to optimize the pump speed.

3 GRADIENT ELUTION DEVICES

In the analysis of complex mixtures containing widely dissimilar components, it is not always possible to obtain a satisfactory separation using one solvent system (isocratic elution). Very often there is insufficient retention of the early emerging compounds and excessive k' values of the late peaks. This situation, known as a *general elution problem,* can be solved by a gradual change in the composition of the mobile phase during the course of the separation. This technique is known as the *gradient elution* or *solvent programming.* The design of the gradient elution devices is determined by the type of pumping system used. Generally, there are two types of gradient elution devices: low-pressure gradient systems and high-pressure gradient systems.

3.1 Low-Pressure Gradient Systems

Low-pressure gradient systems are most often used with a low-volume reciprocating piston or diaphragm pump. The solvents are mixed at atmospheric pressure and the mixture is pumped by a single high-pressure pump. Several low-pressure gradient systems of different complexity are illustrated in Fig. 16. With these systems, the shape of the gradient profile is determined by the volume of liquid in the mixing chamber or reservoir and by the rate at which modifying solvents are added.

Besides being relatively inexpensive, low-pressure gradient systems offer great versatility because several modifying solvents can be handled at the same time. Disadvantages include potential difficulties in operation, reproducibility, and the speed in changing the solvent composition.

3.2 High-Pressure Gradient Systems

These systems are incorporated in most liquid chromatographs, and usually only two-solvent gradients can be generated. In addition to a continuous gradient, these devices are also capable of generating mixed solvents of different composition for isocratic operation. Although designs

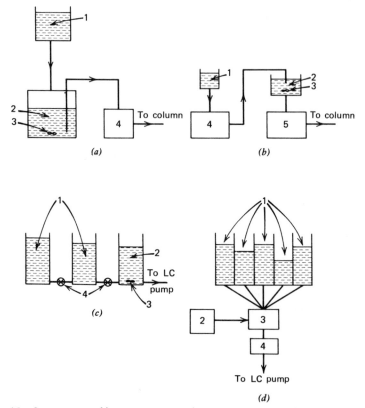

Figure 16 Some types of low-pressure gradient systems. *(a)* As liquid is drawn into the pump, an equal volume of modifying solvent enters the reservoir holding the mobile phase. (1), Modifying solvent; (2), starting solvent; (3), stirrer; (4), pump (high pressure). *(b)* Modifying solvent is transferred to the mobile phase with a second pump. (1), Modifying solvent; (2), starting solvent; (3), stirrer; (4), transfer pump; (5), pump (high pressure). *(c)* Multiple reservoirs containing different solvents permit complex gradient profiles to be produced. (1), Modifying solvents (many possible); (2), starting solvent; (3), stirrer; (4), valves. *(d)* Apparatus for incremental gradient elution. (1), Reservoirs of different solvents; (2), programmer; (3), multiport valve; (4), dilution and mixing volume. Reproduced from reference 22 with permission.

employing a single pump (pneumatic amplifier pump) are available, usually two pumps are employed, each one operating at a fraction of the desired flow rate. By progressively decreasing the delivery of one pump and increasing the delivery of the other, gradients are generated and the two high-pressure streams of solvents are then mixed in a mixing chamber, immediately prior to entering the column. Both the hydraulic units in which the solvents are mixed and the electronic systems for generating different types of gradients have been discussed in the literature. Figure 17 illustrates a gradient system that employs dual-head, special-drive reciprocating pumps.

Generally, with high-pressure gradient systems, three different gradient profiles can be obtained if the initial eluent is a low-concentration solvent: linear, convex, and concave. Concave curves can be described by a general equation:

$$B(\%) = Kt^n \tag{43}$$

and convex curves by:

$$B(\%) = K[1 - (1 - t)^n] \tag{44}$$

where $B(\%)$ refers to the concentration of the second eluent (B), t is the ratio of the time elapsed to the total gradient time, n is a whole number, and K is a proportionality constant. Many different curves are obtainable.

Stepwise gradients, which can generate a large number of complex curves of evenly spaced segments, are also available. The initial and final compositions are selected, and the intermediate compositions are divided into segments of constant duration.

4 SAMPLE INTRODUCTION DEVICES

Generally, four types of devices are used for sample introduction: septum-type syringe injectors, septumless syringe injectors, valve-type injection systems, and automatic sample injectors.

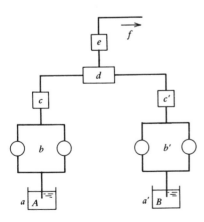

Figure 17 Waters Model 660 solvent programmer. a,a', Solvent reservoirs; b,b', dual-head pumps; c,c', pressure sensors; d, solvent programming manifold; e, high-pressure noise filter; f, to column. Reproduced from reference 23 with permission.

4.1   Septum-Type Syringe Injectors

This is the simplest and the cheapest method of sample introduction. Samples are injected at the top of the column through a self-sealing septum (PTFE-faced septa or fluorinated elastomer materials) using a syringe (Fig. 18). The practical difficulties associated with this method of sample introduction result from the injection method rather than the injector design. The disadvantages of the introduction of sample directly into the column packing have already been mentioned.

On-stream injections are usually the preferred method, since disturbances in the flow streamline may cause deformations in the square-wave profile of the injected sample. On-stream injections can be made at up to 100 atm with single-septum injectors and at up to 200 atm with double-septum arrangements. However, the volume of sample injected is limited due to the force needed to push the plunger against a high back pressure.

4.2   Septumless Syringe Injectors

With injectors of this type, injections are carried out by placing the syringe in an injection block, sealing the needle by compressing the PTFE fer-

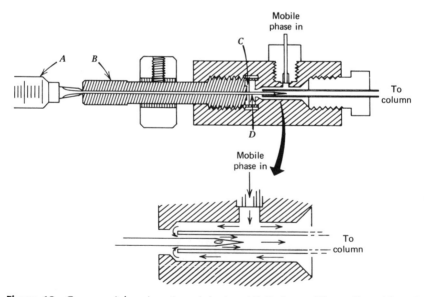

Figure 18   Commercial syringe-type injector. (A) Syringe; (B) needle guide; (C) septum; (D) syringe needle. Reproduced from reference 22 with permission.

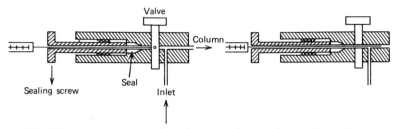

Figure 19 Schematic representation of a septumless on-flow syringe injection port. Reproduced from reference 24 with permission.

rules, and injecting the sample onto the column after the injection port has been opened. These injectors can be operated under high-pressure or stopped-flow conditions. A septumless on-stream syringe injection port is illustrated in Fig. 19.

4.3 Valve-Type Injection Systems

The use of valves of different design has become a very popular way of introducing precise volumes of sample into the chromatographic system. The principle of operation involves injection of the sample at atmospheric pressure into an internal cavity or an external loop, followed by its introduction into the mobile phase stream by switching of the rotator. Usually, valves with internal sample storage permit injections of fixed volumes of sample (1–10 µl). With valves employing external sample loops, larger samples can be injected (8–2000 µl) since loops are detachable. Figure 20 illustrates the principle of operation of an injection valve system capable of introducing variable sample volumes. Valve-type injection devices owe their success to the low solute diffusivity in the solvent held within the loop.

4.4 Automatic Sample Injectors

With increasing popularity of HPLC in quality control and other routine analyses in which operating parameters are predetermined and remain invariant from analysis to analysis, unattended operation of the chromatographic system has become a necessity. Thus microprocessors that are capable of controlling the action of the pumps, variable wavelength detectors, gradient programmers, and automatic injectors, in addition to data handling, have become important in HPLC instrumentation. The automatic injector is a microprocessor-controlled version of the manual

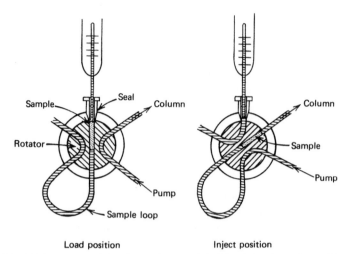

Load position   Inject position

Figure 20 High-pressure injection valve modified for injection of variable sample volumes. Reproduced from reference 24 with permission.

universal injector. Up to 64 sample vials can be loaded into the autoinjector carousel. The system parameters (flow rate, duration of gradient, time between injections, etc.) are chosen, stored in controller memory, and sequentially executed on consecutive injections.

4.5 Band Broadening due to Sample Introduction

The extent of band broadening that is due to sample introduction ($\sigma^2_{v_{\text{inj}}}$) is determined by the method of sampling, the construction of the injection device, and the volume of sample injected (4,7):

$$\sigma^2_{v_{\text{inj}}} = (f v_{\text{inj}})^2 + \sigma^2_{v_0} \tag{45}$$

The proportionality factor, f, is determined by the method of sample injection, the rate at which it is introduced, the flow rate, and the personal skill of the operator. $\sigma^2_{v_0}$ is the variance of the injector extrapolated to an infinitely small volume injected. The variance arising from the sample volume can be considered as a part of the eddy diffusion term in the plate height equation.

If the injection is a true square wave function, the value of the coefficient, f, calculated theoretically, would be 0.288. In practice, higher values are usually encountered (0.4 to 0.5), since ideal injection profiles are rarely realized. For most analytical separations, the maximum sample volume can be calculated from the retention volume of the first peak of interest:

$$v_s = \frac{1}{3} (t_R F_v)$$ (46)

where F_v is the flow rate in volume per unit of time, and t_R is the retention time.

Owing to the dependence of the f parameter on the flow characteristics of the injection device, the injection systems should have minimal and smooth flow paths with no unswept areas. Ideally, samples should be injected directly into the center of the column inlet in a sharp and well-defined spot, allowing the infinite column diameter effect. However, this would require introduction of the sample directly into the packing, which would disturb the bed structure and result in diminished column performance. Alternatively, point injection is accomplished by displacing a sample from a loop to the center of the column inlet.

5 LIQUID CHROMATOGRAPHIC (LC) DETECTORS

The purpose of an LC detector is to "visualize" the separation performed by the chromatographic column, without influencing the extent of the separation. Improperly designed detector cells of large dead volume or with unswept areas will have a deleterious effect on the separation. HPLC detectors are usually subdivided into two categories: *bulk property detectors* and *solute property detectors*.

The bulk property detectors (e.g., refractive index, conductivity, dielectric constant detectors) compare a property of the moving phase with the same property of the solute. The signal (Y) is proportional to the mass flow rate (m/t):

$$Y = K_m \frac{m}{t} = K_m C F_c$$ (47)

where m = total solute mass, t = time, C = concentration, F_c = flow rate, and K_m = proportionality constant.

The area of the solute peak (A) is directly proportional to the total solute mass (m) with no dependence on the flow rate (F_c):

$$A = K_1 m \tag{48}$$

where K_1 is the proportionality constant.

Solute property detectors (e.g., UV absorption, polarographic detector) measure the property of the solute alone. The signal (Y) is proportional to the concentration (C) of solute traversing the detector:

$$Y = K_c C \tag{48a}$$

where K_c is the proportionality constant.

The area of the solute peak is directly proportional to the total solute mass entering the detector, and inversely proportional to the flow rate (F_c):

$$A = K_2 \frac{m}{F_c} \tag{49}$$

where K_2 is the proportionality factor.

Therefore, any variation in the solvent flow will affect the peak area and quantitation with solute property detectors. It should be noted that both Eqs. 47 and 48a are valid only if a linear relationship exists between the detector signal and the sample size.

Regardless of the principle of operation, an ideal LC detector should have the following properties:

1 Low drift and noise level (particularly crucial in trace analysis).
2 High sensitivity.
3 Fast response.
4 Wide linear dynamic range (this simplifies quantitation).
5 Low dead volume (minimal peak broadening).
6 Cell design which eliminates remixing of the separated bands.
7 Insensitivity to changes in type of solvent, flow rate, and temperature.
8 Operational simplicity and reliability.

9 It should be tunable so that detection can be optimized for different compounds.

10 It should be nondestructive.

Some of these factors will be discussed in greater detail in the following paragraphs.

5.1 Drift and Noise Level

Drift is characterized by a continuously rising or falling baseline over a period of time. It is most often associated with change in the mobile phase composition and/or temperature. Drifts that result from the change in eluent composition are sometimes encountered in gradient elution.

Noise is generally defined as the variation in the signal that does not carry information on the parameter to be measured. It can be distinguished as high-frequency or short-term noise and long-term noise.

Short-term noise, also known as "grass," is manifested in high-frequency (greater than 50 Hz) oscillations of the recorder pen. It may be caused by improper grounding of the detector and/or recorder, use of a recorder with too fast a response time, high gain of the recorder amplifier, or random electron movement within the electronics. In most cases, short-term noise can be reduced by means of a capacitance-resistance filter.

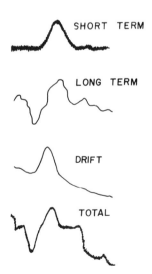

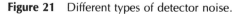

Figure 21   Different types of detector noise.

Long-term noise that results in erratic, random noise is often caused by impurities in the mobile phase (dissolved impurities, stationary phase bleed-off, etc.). Long-term noise of a more regular nature usually results from the malfunctioning of an equipment component or insufficient mixing of the liquids used in gradient elution. Different types of noise are illustrated in Fig. 21.

5.2 Detector Response

The definition of detector response depends on whether it is mass-sensitive or concentration-sensitive. For mass-sensitive detectors, the response (mV/mass/unit time) is given by the following relationship:

$$R_m = \frac{hW}{sM} \tag{50}$$

For the concentration sensitive detector, sensitivity is given in units of mV/mass/unit volume and it can be calculated using the following formula:

$$R_c = \frac{hwF_c}{sM} \tag{51}$$

The symbols in Eqs. 50 and 51 have the following meaning:

h = peak height (mV)
w = peak width at 0.607 of the peak height (cm)
F_c = flow rate (ml/min)
M = mass of solute injected
s = chart speed (cm/min)

In addition to factors such as the nature of the mobile phase, cell geometry, and so on, the detector response is also a function of the type of solute. This is important when comparing two detectors of the same design.

5.3 Detector Sensitivity

Detector sensitivity is one of the most important properties of an LC detector. It is expressed in terms of the concentration of solute (g/ml) or the mass (g/sec) entering the cell per unit time to produce a signal-to-noise (short-term) ratio of 2.

For a mass-sensitive detector, the sensitivity (S_m) is given by the following equation:

$$S_m(\text{g/sec}) = \frac{2N_{tot}}{R_m} \qquad (52)$$

where R_M is the detector response, and N_{tot} is the combined short- and long-term noise.

For a concentration-sensitive detector, the sensitivity (S_c) is given by the following equation:

$$S_c(\text{g/mole}) = \frac{2N_{tot}}{R_c} \qquad (53)$$

where R_c is the detector response.

It should be pointed out that the sensitivity of a detector is not the minimum mass that can be detected. This value is influenced by the chromatographic conditions. Early eluting peaks are usually sharp, whereas the ones with long retention times are broad and sometimes difficult to discern from the noise.

5.4 Detector Dynamic Range

The dynamic range is the concentration range over which the detector will give a concentration-dependent response. The minimum is determined by the detection limit and the maximum by the concentration of solute at which the detector becomes saturated.

In quantitative analysis, it is desirable that the detector response (signal) be directly proportional to the mass or the concentration of the solute (linear dynamic range). This can be tested by plotting the detector response versus the mass of sample injected. Linear behavior is characterized by a straight line with a slope of unity. The relationship between the detector signal and sample size can be expressed as:

$$R = KX^n \tag{54}$$

where R is the detector signal, K is the proportionality constant, X is the concentration or mass rate of the solute, and n is the response index. The latter parameter is a dimensionless constant and it will be unity for a truly linear response. However, in practice it varies between 0.98 and 1.02.

The upper limit of linearity is usually taken as a point at which a 10% deviation is observed. The linear dynamic range is the ratio of the concentration given a 10% deviation (or three standard deviations of the measurements within this range) and the concentration equivalent to the baseline noise. An estimation of the linear dynamic range is particularly important with mixtures containing components of widely different concentrations. For most quantitative work, a linear dynamic range of 10^4 or better is necessary.

Many reviews of the general aspects of LC detection and comparative studies are available in the literature (see Bibliography).

5.5 Detector Contribution to Band Broadening

The combined short- and long-term detector noise, N_{tot}, can be expressed in terms of the maximum amplitude over a period of approximately 10 min:

$$N_{tot} = V_i A \tag{55}$$

where V_i is the amplitude in mV, and A is the attenuation factor.

It has been shown before that the variance of the detected peak is the sum of the variances caused by the column, the injection system, and the detector. If the contribution of the injection system is neglected, and if it is assumed that a tolerable loss in resolution is 5%, the following expression is obtained (39):

$$\sigma_{t_{det}}^2 = 0.1 \, \sigma_{t_{col}}^2 \tag{56}$$

or

$$\sigma_{t_{det}} = 0.32 \, \sigma_{t_{col}} \tag{57}$$

The column variance can be expressed as:

$$\sigma_{t_{col}} = \frac{t_R}{N^{1/2}} = \frac{N^{1/2} (1 + k')}{N} = \frac{N^{1/2} (1 + k')H}{u} \qquad (58)$$

By substituting Eq. 58 into Eq. 57, one obtains:

$$\sigma_{t_{det}} = \frac{0.32N^{1/2} (1 + k')H}{u} \qquad (59)$$

The detector contribution (in volume units) to the total peak width can be described by the following formula (32):

$$\sigma_{v_{det}} = \sigma_{t_{det}}F \qquad (60)$$

where F is the mobile phase flow rate. The requirement for small cell volumes is particularly demanding with highly efficient microparticulate packings which give very narrow bands. The peak broadening due to the dispersion in the detector cell is especially pronounced with early eluting compounds ($k' < 2$). This effect can be minimized by using detectors with cell volumes of $^1/_5$ to $^1/_{10}$ the peak volume. It should also be pointed out that if several detectors are used in series, the first detector should have the smallest cell volume.

5.6 Types of Detectors

In practice, there is no truly universal LC detector that would respond to all compounds with equal sensitivity. This has resulted in the proliferation of detection devices utilizing specific solute properties. Table 5 summarizes the characteristics of some commercially available detectors. In the following section, only the most commonly used LC detection systems will be discussed:

1 Spectrophotometric detector.
2 Fluorescence detector.
3 Refractive index (RI) detector.
4 Electrochemical detector.

Table 5   A Comparison of the Characteristics of Some LC Detectors[a]

DETECTOR PARAMETER	POLARGRAPHIC	CONDUCTIVITY	FLUORESCENCE
Principle of operation	Measurement of the current change between polarizable and non-polarizable electrodes	Electrical conductivity due to the presence of solute in moving phase (differential or absolute)	Adsorption of UV light by solute, which then fluoresces at a higher wavelength (often visible) excitation at 360 nm; emission at 400 – 700 nm)
Solute/solvent limitations	Solvent(s) must be chosen to form a suitable electrolyte for the solute(s)	Solvent(s) must be chosen so that the background conductivity can be compensated without loss in detector performance	Solvent should not fluoresce to any significant extent (solvent signal< 1000 detection limit acceptable)
Temperature coefficient of response	1.5%/°C	2%/°C	Negligible
Detector cell volume	10 µl	1.5 – 2.5 µl	10 – 13 µl
Drift rate	No figure available	No figure available	1 mV h^{-1}on x 1
Noise level	$\pm 1 \times 10^{-10}$ A	$\pm 5 \times 10^{-4}$ µmho	$\pm 0.1 - 0.15$ mV on x 1
Response	$1-2 \times 10^{-2}$ A l mol^{-1}	2.0×10^5 µmho cm^3/g of NaCl in H$_2$O	4×10^8_m , V cm^3/g
Detection limit	2×10^{-8} mol l^{-1} (1×10^{-9} g/cm^3 for Cu^{2+} with 0.5 NaCL in H$_2$O as mobile phase.	0.01 p.p.m. NaCl in H$_2$O ($\sim 1 \times 10^{-8}$ g/cm^3)	1×10^{-9} µg/cm^3 of quinine sulphate
Linear range	1×10^5	1×10^6 with zero suppression	6.4×10^3

DETECTOR PARAMETER	POLARGRAPHIC	CONDUCTIVITY	FLUORESCENCE
Operating Range	0.2-100 A (Ref. 6)	1-1000 mho FSD in 7 steps and recorder attenuation of 0.1-12.8 (binary) i.e. 1 x 0.1 = 0.1 mho FSD	
Reproducibility	$\pm$ 1%	$\pm$ 1%	about 1%
Temperature control	No figure available	$\pm$ 0.005°C for 1% on 0.1 mho FSD. If required, temperature compensation can also be applied as a suitable alternative	not generally required
Gradient elution potential	Severely restricted	Limited use	limited use
Principle of operation	Either absolute or differential absorbance between pure solvent and solvent + solute at 254, 280 or variable wavelength 190-800 nm	Measurement of difference in refractive index between solvent and solvent + solute	Removal of solvent, total compustion of solute followed by reduction to CH_4 and detection by FID (ion current changes)
Solute/solvent limitations	Performance improves with increasing difference in absorbance between solute and solvent	Performance improves with greater difference in refractive index between solute and solvent	Boiling point differential between solute and solvent of 50-100°C. Involatile buffers
Temperature coefficient of response		about 1×10^{-4} RIU/°C (Ref. 3)	FID-$0.1\%/°C$; coating $-0.22\%/°C(H_2O$; $0.35\%/°C$ (CCl_4)
Detector cell volume	Normally 8-25 µl	3-12	Indeterminate for 'real' cell volume
Drift Rate	0.0005 ODU/h without temperature control	1×10^{-6} RIU/h without temperature control; 2×10^{-5} RIU/h with temperature control	2×10^{-3} A h^{-1}

Table 5 (*Continued*)

DETECTOR PARAMETER	POLARGRAPHIC	CONDUCTIVITY	FLUROESCENCE
Noise Level	$\pm$0.0001 ODU	1×10^{-7} RIU	$\pm 1 \times 10^{-13}$ A for complete system
Response	4×10^5 ODU cm^3 g^{-1}	22 RIU cm^{3}g^{-1}	Effective response -2×10^{-5} c g^{-1} (assume 2×10^{-2} c g^{-1} for FID and 0.1% pick-up on wire)
Detection limit	1×10^{-9} g cm^{-3} phenonthrene in hexane	9×10^{-9} g cm^{-3} sucrose in water	$1-5 \times 10^{-6}$ g cm^{-3}, solute unspecified
Linear range	5×10^3	5×10^3 for deflection and Fresnel designs	FID-1 $\times 10^7$; amplifier 1×10^5; system $\sim 1 \times 10^3$
Operating range	$0.01 - 2.54$ ODU FSD	1×10^{-7} -5.12×10^{-3} RIU FSD	x1-x1024 (binary) x1 $-$x 10^3 (decade) i 1×10^{-12} A $- 1.024 \times 10^{-6}$ A FSD
Reproducibility	$\pm$ 1%	$\pm$ 1%	better than $\pm$ 3%
Temperature control	Not generally required	$+$ 0.01°C to control within $\pm$ 1 $\times 10^{-7}$ RIU	$\pm$ 0.1°C; except coating block at ambient
Gradient elution potential	Limited use (non-UV adsorbing solvents)	Strictly limited use	Wide applicability - except nonvolatil buffers

[a]Reproduced (with modifications) from reference 26 with permission.

5 Phase transformation detector.

6 Radioactivity detector.

7 Dielectric constant detector.

8 Electrical conductivity detector.

9 Heat of adsorption detector.

10 LC-mass spectroscopy (MS) coupling.

11 Reaction detectors.

5.6.1 Spectrophotometric Detectors

The operation of this type of detector is based on the measurement of absorbance of UV or visible light. Due to its high sensitivity, wide dynamic range, and relative insensitivity to temperature and flow variations, the spectrophotometric detector has become one of the most widely used LC detectors. There are four general types of spectrophotometric detectors:

1 Single wavelength detectors, employing a 254-nm mercury line.

2 Multiwavelength detectors that operate at several (most often 2) discrete wavelengths and employ narrow band emission lines from discharge lamps or from luminescent coatings on the UV source.

3 Continuously variable wavelength detectors that enable selection of detection wavelength by means of a monochromator. The ability to vary the wavelength of detection is advantageous in maximizing the sensitivity (measurements in the vicinity of λ_{max}) and minimizing the background absorbance of the mobile phase when necessary. In addition, interferences from other compounds with different absorption characteristics can be minimized or eliminated.

4 UV-visible LC spectrophotometers equipped with stopped-flow scanning capabilities (wavelength scanning drive and memory module for background compensation) with which complete UV-visible spectra can be obtained. Scanning is usually performed at a speed of 100 nm/min. The background spectrum of the solvent is stored in the memory module and later subtracted from the spectrum of the peak arrested in the cell. It should be pointed out that because of low solute diffusivity in the mobile phase, the flow can be stopped for several hours without any significant loss of the solute from the detector cell.

Rapid-scanning spectrophotometers are currently being developed for on-line use with the liquid chromatograph. Spectra can be obtained within a fraction of a second, which is short compared with the peak width. These spectrophotometers employ a moving radiation source, such as a radiation point at a cathode-ray tube or a rotating mirror in the light path of the monochromator. Regardless of the scanning method used, the UV or visible spectra provide important information concerning the identity of the peaks under study.

Superimposed on this division are two additional modes of operation of spectrophotometric detectors, depending on whether a reference path is provided or not. Thus single-beam (no reference path) and double-beam (with reference path) spectrophotometers are commercially available. Optical layouts of the two types of detectors are shown in Fig. 22. The double-beam instrument is generally preferred in most chromato-

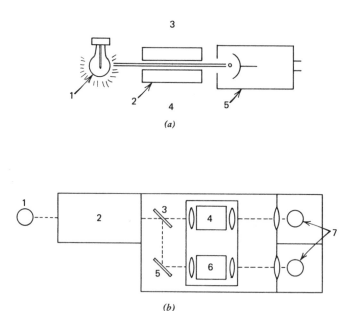

(a)

(b)

Figure 22 Optical layout of single- and double-beam photometric detectors. (a) Single-beam detector, illustrated in the form of a fixed-wavelength photometer. (1), Spectral source, for example, low-pressure mercury lamp; (2), flow cell; (3), outlet; (4), inlet; (5), phototube. (b) Double-beam detector, illustrated in the form of a variable-wavelength photometer. (1), Spectral source, for example, deuterium lamp; (2), monochrometer; (3), beam splitter; (4), analytical flow cell; (5), mirror; (6), reference flow cell; (7), photodiodes. Reproduced from reference 22 with permission.

graphic work, since the reference light beam compensates for fluctuations in the light source and variations arising from effects other than absorption.

The most commonly used spectral source is a low-pressure mercury lamp that emits light of high intensity at a wavelength of 253.7 nm. At this wavelength, most organic compounds absorb strongly, whereas the commonly used mobile phases do not have a high background absorbance. "Stray" emission can be eliminated by means of a narrow bandpass interference filter. Hydrogen or deuterium discharge lamps operated under low pressure and DC conditions are also used. They provide continuous UV emission down to 165 nm. At wavelengths longer than 360 nm, emission lines are superimposed on a continuum. Therefore, measurements above this wavelength are made using the tungsten lamp.

The principle of operation of a spectrophotometer is based on the fundamental law of spectrophotometry, the Beer-Lambert law, which states that the fraction of radiation absorbed is proportional to the number of absorbing species:

$$I = I_0 e^{a'bc} \tag{61}$$

where I = intensity of the transmitted beam, I_0 is intensity of the incident beam, a' is a constant known as the absorptivity, b is the radiation path length, and c is the concentration of the absorbing species. The absorptivity depends on the wavelength of the radiation and the nature of the absorber. The product of absorptivity and formula weight is called "molar absorptivity," ε.

By rearranging Eq. 61 and taking a logarithm, one obtains:

$$\log \frac{I_0}{I} = abc \tag{62}$$

The left-hand side of the equation is termed absorbance, A, so that Eq. 62 becomes:

$$A = abc \tag{63}$$

Due to secondary processes such as reflection, scattering, and absorption by the cell walls, even in the absence of absorbing species, there is a slight decrease in radiant power of the incident beam in both the sample and the reference path:

$$I_{0\,sample} \rightarrow \text{sample cell} \rightarrow I_1 \tag{64}$$

$$I_{0reference} \rightarrow \text{reference cell} \rightarrow I_2 \tag{65}$$

Since compensations are made so that

$$I_{0sample} = I_{0reference} \tag{66}$$

the relationship for the sample beam becomes:

$$\log \frac{I_0}{I_1} = \text{absorbance of sample } + \text{ absorbance of solvent} \tag{67}$$

and for the reference beam:

$$\log \frac{I_0}{I_2} = \text{absorbance of solvent} \tag{68}$$

The difference between the two gives the absorbance due to the sample, or

$$\log \frac{I_1}{I_2} = abc \tag{69}$$

In practice, when working with solvents of high transmittance (greater than 75%), the reference is usually air. Conversely, with solvents that absorb at the operating wavelength, some compensation of the change in background absorbance is needed. In isocratic operation this can be accounted for by filling the reference cell with a constant-composition mobile phase. In gradient elution, however, the mobile phase absorbance has to be corrected for electronically (by means of a memory module) or by splitting the mobile phase immediately before the sample injector. This requires either an electronically controlled delay or the use of a dummy (reference) column closely matched with the analytical column. However, in most gradient elution applications, it is possible to avoid baseline compensation by careful selection of solvents of desired properties.

Several cell geometries are in common usage: the Z-type cell, the H-type cell, and the Tapered cell® (Fig. 23). A well-designed cell should not be subject to refractive index (RI) effects, it should contain no unswept areas, and it should be able to withstand pressure. Effects of flow and/or refractive index variations can be diminished by proper cell design or thermostating of the cell and eluent by means of heat exchangers. Most currently used cells have volumes of 8 to 10 μl and an optical path length

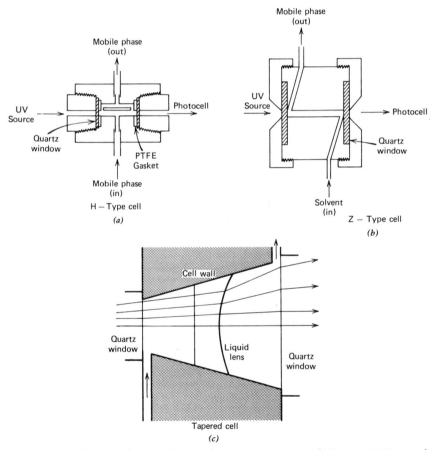

Figure 23 Different cell design for UV detectors; (a) H type; (b) Z type; (c), Tapered cell. Figures 23a and b are reproduced from reference 26 with permission. Figure 23c is reproduced from reference 30 with permission, of Waters Assoc. Inc.

of 10 mm. With the recent trend toward 5-μm packing materials and microbore columns, new cell designs and cell volumes of approximately 1 μl are being investigated in order to minimize peak broadening and increase sensitivity.

5.6.2 Fluorescence Detectors

Fluorometric detection is becoming increasingly popular due to its high sensitivity and selectivity. Its use has been particularly extensive in biochemical and biomedical work, since many biologically important compounds such as porphyrins, riboflavins, aflatoxins, vitamins, certain drugs, and drug metabolites, and so on, fluoresce strongly. Thus it is possible to monitor a few selected compounds in a complex biological matrix sensitively and free from interferences.

In principle, the eluting compounds are irradiated with light (excitation) from an intense source (usually UV) and the light emitted is measured. Except in the case of resonance fluorescence, the radiation emitted is usually of longer wavelength. Thus both the excitation and emission wavelength are informative. The optical layout of a fluorometer is shown in Fig. 24.

The excitation wavelength is selected by using a monochromator. With respect to the optical arrangement, fluorometers can be constructed with a photomultiplier in line with the radiation source or at a right angle to it. In addition, both single- and dual-flow cell arrangements are available commercially. The latter arrangement compensates for the fluorescence of the mobile phase.

Fluorometric detection possesses inherently higher sensitivity than the absorption methods because the intensity of fluorescent emission is directly proportional to the power of the exciting radiation. Picogram amounts of highly fluorescent compounds can be detected routinely. However, at high solute concentrations, strong absorption occurs and the intensity of the source is effectively decreased; thus there is apparent nonlinearity of the detector response. Therefore, quantitative work should be performed with dilute samples. Several factors may affect the linear range in fluorescence monitoring:

1 Overloading of the photoreceptor or amplifier.
2 External conversion: energy transfer between the excited molecules and the solvent or other solutes.

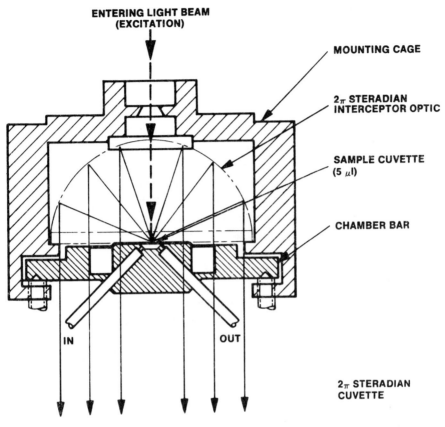

Figure 24 Cross-sectional view depicting the essential components of the flow-through curvette of the fluorometer. Courtesy of Kratos, Inc., Schoeffel Instrument Corporation.

3 Internal conversion: radiationless deactivation of the excited molecules.

4 Predissociation: rupture of the bonds of the absorber due to the high vibrational energy of the lower energy state.

It should also be pointed out that special demands are placed on the purity of the solvents used in fluorescence measurements, since considerable quenching can be caused by the impurities and dissolved oxygen. In addition, temperature and solvent viscosity can influence fluorescence emission. Low temperatures and high solvent viscosities are desirable since they reduce radiationless deactivations of the excited solute molecules. Unfortunately, the temperature range over which LC detectors are operated is quite narrow, and the number of solvents useful for RP systems is limited.

Compounds that do not fluoresce naturally can be derivatized by pre- or postchromatographic (between the column outlet and the detector cell) reactions. Derivatization of amines and alcohols with fluorescamine or dansyl chloride have been successfully carried out, and many examples are reported in the literature (31). By careful choice of specific fluorescence-labeling reactions, increased sensitivity and selectivity can be achieved. Unlike absorption and RI measurements, fluorescence signals cannot be quantitatively expressed in terms of a universal solute property. Therefore, a reference compound is needed and most often solutions of quinine sulfate in 0.1-F sulfuric acid are used.

5.6.3 Refractive Index (RI) Detectors

The operation of this mass-sensitive detector is based on the measurement of the difference in refractive indices of the liquids contained in the sample and reference cells. This detector is very versatile, and due to its nonspecificity, it nearly approximates the universal LC detector. However, owing to the relatively small differences in the absolute refractive indices of many substances commonly analyzed by HPLC, the sensitivity of these devices is generally lower than that obtained with UV and fluorescent detectors (usually between 10^{-6} to 10^{-8} g/ml of column effluent). The limit of detection also depends on temperature, sample type, and the nature of the mobile phase.

The marked dependence of refractive indices on temperature makes thermostating of both cells mandatory in order to eliminate temperature differences between the two flow cells. Therefore, commercial differential RI detectors incorporate heat exchangers of carefully controlled design to avoid additional peak broadening. Since the temperature coefficient of the RI detector is 1.10×10^{-4} RIU/C°, a temperature control of ± 0.001°C is necessary if noise level is to be maintained at 1×10^{-7} RIU.

Since the detector operation is also sensitive to pressure variations, a pulseless flow is essential. In addition, the detector response is affected by variations in the mobile phase composition, and gradient elution is not possible unless the detector is operated in the truly differential mode, that is, with two columns and two flowing liquid streams. Even with this operation, small differences in pressure between the two cells can cause an excessive drift in the baseline.

The linear range of the RI detector depends on its principle of operation. The three most widely used RI detectors are the Fresnel type, the deflection type (or differential refractometer), and the interferometer.

The Fresnel-type RI detector

In the Fresnel-type RI detector (Fig. 25), the light fraction reflected at a dielectric interface (glass-liquid) is related to the refractive index and the angle of incidence. For maximum sensitivity, the light is directed at an angle that approximates the critical angle. The intensity of the beam is measured after its reflection by the metal surface at the back wall of the cell. This compensates for the refractive index of the mobile phase. Alternatively, a matched cell containing a reference liquid is used to compensate for the mobile phase effects and variations in temperature and the source output. With the latter operation, a differential photodetector is necessary. In the Fresnel-type detector, the cell volumes are usually small (approximately 3 μl), and therefore they are ideally suited

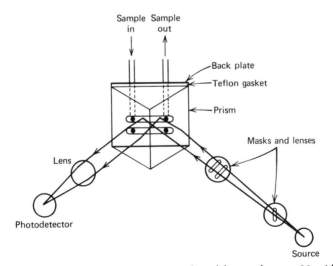

Figure 25 Fresnel-type refractometer. Reproduced from reference 22 with permission.

for work with high-efficiency columns. Although these cells do not contain unswept areas, they are, nevertheless, susceptible to formation of solid films, which give rise to cell imbalance and require careful and delicate cleaning. For preparative work, larger versions are also available. A common disadvantage of the Fresnel detector is that two prisms are needed to cover the useful RI range (1.31 to 1.55).

The deflection-type RI detector

With the deflection-type RI detector, the difference in the refractive indices of the sample and reference liquids is detected through deflection of light passing through the sample and reference cells, separated by a diagonal sheet of glass (Fig. 26). The light beam is transmitted through the dual cells, reflected by a mirror, and passed back through the cell a second time. This type of detector needs only one cell to cover the whole RI range. Although the cells are less susceptible to contamination than the Fresnel type, they are, however, less efficiently swept.

Interferometer

Interferometers employ the Christiansen effect as the basis of their operation, that is, the dependence of the transmission of light through a heterogeneous medium on the difference in the refractive indices of the phases. This detector is shown diagrammatically in Fig. 27. Due to the difference in the RI values between the sample and the reference, the two emerging beams are out of phase. Interference of the two beams combined by a birefringent element is measured. Characteristics of the interferometer are available in the literature (22).

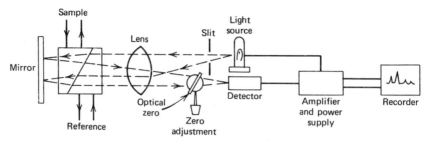

Figure 26 Schematic diagram of a refractometer. Reproduced from reference 22 with permission.

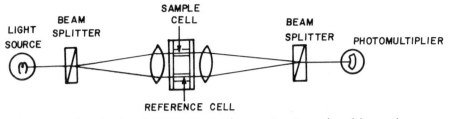

Figure 27 Shearing interferometer-type refractometer. Reproduced from reference 30 with permission.

5.6.4 Electrochemical Detectors

The use of hydrodynamic thin-layer electrochemistry for detecting electroactive compounds in an HPLC eluent is rather recent. The development of electrochemical detectors has been prompted by the insufficient sensitivity of most commonly used LC detectors for trace analysis of endogenous levels of biologically important compounds. Complex biological matrices, such as urine, are known to contain a large number (over a 100 UV-absorbing compounds alone) of components of widely different concentrations. Since procedures for preconcentration of trace compounds from large samples are rarely very selective, this results in an overabundance of secondary compounds that can overload the system and obliterate the peaks of interest.

The electrochemical detectors possess several distinct features:

1 High sensitivity.
2 Selectivity.
3 Compatibility with the RP systems.
4 Wide linear range.

Since picogram and even femtogram levels of electroactive compounds can be detected with most commercially available electrochemical detectors, their use has opened up new areas of chemical research, difficult or impossible to deal with thus far. The detection limit, determined by the signal-to-noise ratio, depends on the rate of electrochemical reactions (coulometric yield) and the level of residual current. The former factor depends on the diffusion coefficient of the solute, the area of the electrode, and the flow rate. The residual current is determined by the magnitude, sign, and duration of the applied DC voltage, pressure variations in the flow, and the presence of impurities in the solvent system.

Since the number of electroactive compounds within a given potential range of solvent decomposition is low, the selectivity afforded by the electrochemical detector offers considerable advantages over other methods of detection. By varying the polarizing voltage and/or eluent composition, the electrochemical reactivity of functional groups can be affected markedly.

Due to the need for an electrically conducting medium, the choice of solvents compatible with the electroanalysis is limited. Aqueous buffers and their mixtures with organic modifiers such as acetonitrile and methanol used in RP operation afford high conductivity and lend themselves well to electroanalytical work.

Analytical variations in electrochemical detection can be carried out by two different techniques:

1 *Voltametric.* The DC voltage applied to an electrode varies steadily over a period of time and the resulting current is measured. If a dropping mercury electrode is used, the process is known as "polarography."

2 *Amperometric.* The potential is held at a constant DC value and the current is measured. If the conditions are controlled so that all active material reacts at the electrode surface, the process is termed "coulometric analysis."

Regardless of the technique used, the current through the interface of the working electrode is always measured. Usually, a three-electrode potentiostatic system is employed, that is, a working electrode, current-carrying auxiliary electrode, and a zero-current reference electrode (Fig. 28). An electronic feedback loop maintains the potential of the reference electrode at a given value. The working electrode can be either a dropping mercury electrode (DME) or a solid electrode made of carbon paste (suspension of carbon powder in mineral oil), glassy carbon, or platinum. The advantages of DME in electroanalytical work are well known. Its obvious disadvantage is in the relatively low anodic range (approximately +0.3 V versus SCE) and high charging current.

With solid electrodes, adsorption of solutes and aging of the surface

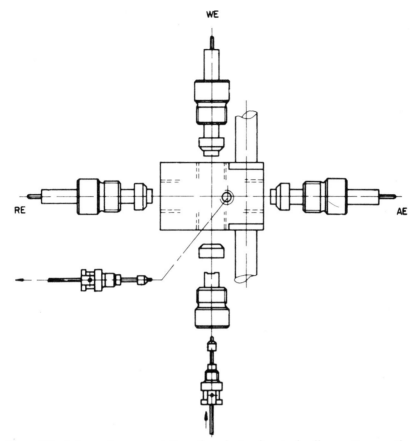

Figure 28 Schematic representation of an electrochemical cell operating on a three-electrode potentiostatic system: working and auxiliary electrodes; glassy carbon or carbon paste; reference electrode; silver-silver chloride. Courtesy of METROHM.

may occur, leading to decreased reaction rates and diminished sensitivity. Although the cathodic range of these electrodes is smaller than that of DME, the anodic range extends to over $+1.5$ V versus SCE.

5.6.5 Phase Transformation Detectors

The concept of the phase transformation detector was borrowed from GC, and the flame ionization detector (FID), one of the most widely used detectors, has been adopted for use in LC. If the mobile phase is volatile and if it does not leave any residue upon evaporation, it can be removed from the column effluent, leaving the solute coated on a moving belt, wire, or chain. The nonvolatile residue consisting of sample components is then carried to a high-temperature oxidation (or pyrolysis) chamber. The gaseous products are fed into a FID detector directly or after catalytic reduction of CO_2 to CH_4. Due to the multistage nature of the phase transformation process, the system is rather bulky. A schematic representation is given in Fig. 29. The sensitivity of this detector is limited by the effectiveness of the transport system, that is, by the fraction of the total eluent transported to the detector. Memory effects and fatigue of the metal used for the solute transport can give rise to noise. In addition, because of the high sensitivity and operational complexity, the system is susceptible to contamination. However, the detector is universal and its response is directly proportional to the carbon content of the sample.

5.6.6 Radioactivity Detectors

Highly sensitive radioactivity measurements have long been used in studies of drug metabolism, pesticide analysis, and so on. Usually, individual

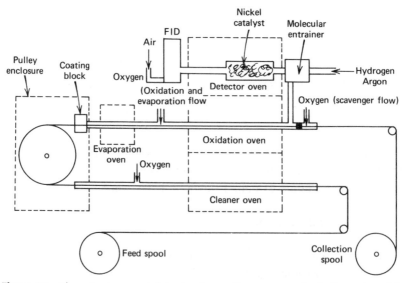

Figure 29 The wire transport detector (by methane conversion). Reproduced from reference 28 with permission.

samples from a fraction collector are counted using the Geiger or scintillation counting systems. This has prompted researchers to adapt the technique for on-line detection of HPLC effluents. Various designs of flow-through radioactivity monitors are available. The general problem encountered with radioactivity detectors stems from the basic requirements for long residence time of the radioactive substance in the detector cell for highest sensitivity. This is contrary to the HPLC requirements for short residence time of the eluent in the detector cell to achieve the most efficient chromatographic separations. A compromise can be reached by increasing the counting time by use of lower flow rates or larger sample cells. However, due to the low rate of solute diffusion in the mobile phase, counting can be performed under stopped-flow conditions with little loss of efficiency.

5.6.7 The Dielectric Constant Detector

The dielectric constant detector measures the change in capacity of a condenser upon passage of the column effluent between its plates. The major disadvantage of this detection system is its incompatibility with gradient elution, flow programming, or temperature programming. In addition, the response is linear only over a narrow concentration range.

5.6.8 The Electrical Conductivity Detector

Continuous measurement of the change in electrical conductivity of the effluent is a sensitive detection method for both inorganic and organic ionic species. The system usually employs two horizontal platinum electrodes which are positioned at the column outlet and form one arm of an AC bridge. The detector response is linear over a narrow concentration range. In addition, the operation of the detector is incompatible with gradient elution, unless the conductivity of the mobile phase remains constant during the course of the gradient.

5.6.9 The Heat of Adsorption Detector

The heat of adsorption detector consists of a flow-through cell containing a small plug of an adsorbent (usually activated charcoal or silica gel) and a thermocouple or a thermistor embedded in the adsorbent or positioned downstream from it. The heat-sensitive element continuously measures the temperature change in the adsorbent or in the mobile phase resulting from adsorption-desorption processes of the solute molecules. If the cross-sectional geometries of the detector and the column are the same, and if the detector is packed with the same adsorbent as the column, the output from the detector will have a differential form of the Gaussisan curve. This detection device has not yet achieved widespread use due to its incompatibility with gradient elution, temperature programming, and periodical contamination of the adsorbent.

5.6.10   Liquid Chromatography–Mass Spectrometer (LC-MS) Systems

The coupling of the gas chromatograph (GC) with a mass spectrometer (MS) gives an unsurpassed combination of separating power and positive identification of chromatographic peaks. Thus the development of an analogous system with LC was a logical step. The off-line use of LC with MS has occurred for quite some time: the column effluent is collected by means of an automatic fraction collector and the fractions are evaporated to a small volume (approximately 20 μl) and analyzed by MS. The great advantage of MS is that it can provide elemental analysis, molecular weight data, and identification of molecular fragments using a minimal sample size. Although the sensitivity of MS is determined by the type of instrument and the nature of the sample, usually a 5 to 10-ng sample will suffice for positive identification. The minimal sample size requirements may vary since some samples contain a high background matrix which can interfere with the identification of the components of interest. In addition, peaks with long retention times from the previous separation may be slowly released from the column, thus obscuring the identification.

However, a major problem in the development of on-line coupling is the need for low operational pressures of the MS (approximately 10^{-6} mm Hg) (1–12). Therefore, its ability to handle liquids is rather limited, and special coupling systems have to be designed to interface the two instruments. There are two coupling schemes presently used for on-line LC-MS system, and these are classified by the method of sample introduction: direct inlet sampling and separation of solvent before ionization. Each interface system has advantages and disadvantages, depending on the type of application. However, regardless of the type of interface used, the system should have the following characteristics:

1 It should not limit the operating conditions of the LC system.
2 It should provide a relatively high solute transfer (>30%) in a reproducible manner.
3 Sample components should not undergo decomposition during vaporization.
4 Separated chromatographic bands should not be significantly broadened by the interface.

Direct inlet sampling

In this approach, developed by McLafferty and his co-workers (7), a portion of the column effluent is fed directly into the MS and the vaporized mobile phase is used as the chemical ionization (CI) agent. The interface is shown in Fig. 30. With gradient elution, the continuously changing solvent composition is accompanied by a concomitant change in the nature of the ionizing agent, which makes the interpretation of CI spectra difficult. Furthermore, the range of useful solvents is restricted even with isocratic operation. The success of the analysis depends on the vapor pressures of solutes at normal MS temperatures. Thus a large number of compounds typically analyzed by HPLC are excluded. Finally, since the CI spectra consist mainly of $M^+ + 1$ ions and the degree of fragmentation is low, they provide limited information.

Direct coupling without solvent separation before the ionization step can also be achieved by introduction of the effluent through a capillary

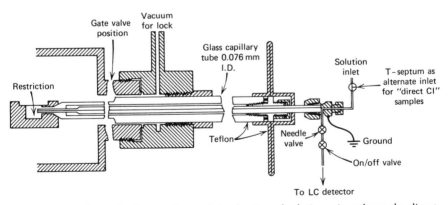

Figure 30 Inlet probe for continuous introduction of solutions; it replaces the direct introduction sample probe of the RMH-2, utilizing the same vacuum lock system. Reproduced from reference 28 with permission.

leak, which is immersed in the liquid at one end and connected to the MS by the vacuum system at the other end.

Horning et al. (13) proposed the "atmospheric pressure ionization" (API), in which the column effluent is mixed with a hot carrier gas (N_2) and vaporized outside of the MS. The vaporized mixture is then passed through an external ionization source (corona discharge or radiation from ^{63}Ni) and into the quadrupole mass analyzer. The sensitivity of the API interface system is similar to that of the CI method.

The wire transport system

An alternative method, developed by Scott et al. (18), employs a moving wire system to transport the solute into the ion source of a quadrupole MS (Fig. 31). The transport system is similar to the one used in phase transformation detectors. The design of the interface is critical, since there must be continuous passage of the wire from atmospheric pressure to the ion source (10^{-6} mm Hg) and out again. This system can be used with any solvent, and it is compatible with gradient elution. Furthermore, the impact spectra are more informative than the CI spectra. Since the mobile phase is completely removed prior to passage of the wire into the ionization chamber, the impact spectra are independent of the nature of the solvent used in the separation.

Further improvement of this system was achieved by McFadden and co-workers (19). They tried to improve the low sample utilization (approximately 1%) resulting from the limited wettability of the wire transport system. By using a stainless steel ribbon, more efficient sample transport was obtained (30 to 50%).

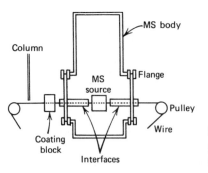

Figure 31 Principle of LC-MS coupling as proposed by Scott et al. Reproduced from reference 24 with permission.

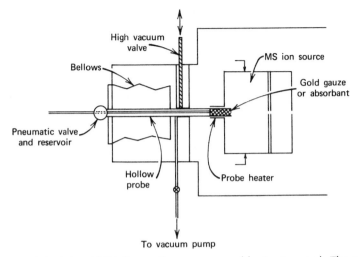

Figure 32 Principle of LC-MS coupling as proposed by Lovins et al. The position of the probe tip during the recording of the mass spectra is shown. Reproduced from reference 24 with permission.

An alternative design, developed by Lovins et al. (20), employs a UV detector as a part of the interface between LC and MS. The purpose of the UV detector is to indicate when the substance of interest has eluted so that the corresponding eluate fraction can be collected and sucked into the volatilization chamber, where the mobile phase is flash evaporated. The solute remaining is then transferred into the ion source by means of a high vacuum system. This type of coupling is illustrated in Fig. 32.

Jones and Yang (21) have proposed an interface system that involves only partial removal of the solvent. A schematic diagram of their system is given in Fig. 33. A three-stage silicone-rubber molecular separator transmits nonpolar solutes while rejecting polar solvents. Since the maximal working temperature is 250°C, only solutes that have an appreciable vapor pressure at this temperature can be analyzed. In addition, the membrane separator cannot be used with nonpolar solvents.

Many additional designs have been described in the literature. In spite of the variety of available coupling schemes, the on-line operation of LC-MS is not yet completely perfected for routine use. The difference in the dynamic behavior of the two techniques results in problems associated with the handling of large solvent volumes and an inability to analyze substances of limited volatility. However, with future improvements, this combination of the powerful techniques will realize its full potential.

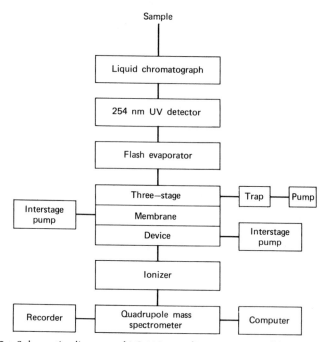

Figure 33 Schematic diagram of LC-MS coupling as proposed by Jones and Yang. Reproduced from reference 24 with permission.

5.6.11 Reaction Detectors

If there is a need for increasing detection sensitivity or reducing interference from closely eluting or overlapping bands, postcolumn reaction or derivatization of the bands can be utilized. Alternatively, column effluents can be illuminated by an intense beam of light to initiate photochemical reactions. Chemically modified or derivatized compounds are then detected on-line, either spectrophotometrically or fluorometrically. Although this idea has been classically employed in amino-acid analyzers (the ninhydrin reaction), its use in HPLC is recent. The application of reaction detectors in HPLC requires careful design of the reaction-detector system, because extracolumn effects can cause considerable band broadening. Since most derivatization reactions, outlined in Table 6, take place in aqueous buffers or solvents miscible with water, this detection technique is ideally suited for RP analyses. The major requirements for

Table 6 Some Derivatization Reactions Used in Reaction Detectors

Sample	Derivatizing Agent	Detection	Reaction Product λ_{max}
Carboxylic acids	p-Bromophenacyl bromide	Photometric	260
	O-p-Nitrobenzyl-N,N'-diisopropylisourea (II)	Photometric	265
	1-(p-Nitro) benzyl-3-p-tolyltriazine	Photometric	265
Amino Acids	Pyridoxal	Fluorometric	330 (excit.), 400 (emiss.)
Amino Acids, Amines, Phenols, Peptides	Dansyl chloride (5-N,N-dimethylaminoaphthalene-1-sulfonyl chloride)	Fluorometric	360 (excit.), 510 (emiss.)
Amines	p-Methoxybenzoyl chloride	Photometric	254
	2,4-Dinitro-1-fluorobenzene	Photometric	360
	N-Succinimidyl-p-nitrophenyl acetate	Photometric	265
Alcohols	Benzoyl chloride	Photometric	230
	p-Nitrobenzoyl chloride	Photometric	254
Aldehydes, Ketones	p-Nitrobenzyloxyamine·HCl	Photometric	265
	2,4-Dinitrophenylhydrazine	Photometric	430
1,2-, 1,3-, 1,4-Diols	Phenanthrene boronic acid	Fluorometric	313 (excit.), 385 (emiss.)

Table 6 Some Derivatization Reactions Used in Reaction Detectors (*continued*)

Sample	Derivatizing Agent	Detection
Amines, Amino Acids	Fluorescamine (Fluram)	Fluorometric, 390 nm (excit.), 475 nm (emiss.)
	o-phthaldehyde	Fluorometric, 340 nm (excit.), 455 nm (emiss.)
	Ninhydrin	Photometric, 440 nm, 570 nm
Peptides	Fluorescamine	Fluorometric, 390 (excit.), 475 nm (emiss.)
Carboxylic Acids	o-Nitrophenol, Sodium salt	Photometric, 432 nm
Aldehydes, Ketones	2,4-Dinitrophenylhydrazine	Photometric, 430 nm
Enzymes	Substrates	Photometric/Fluorometric
Substrates	Enzymes	Photometric/Fluorometric
Phenols, Carbohydrates, Carboxylic Acids	CE^{4+}	Fluorometric; 260 nm (excit.), 350 nm (emiss.)
Catecholamines	Hexacyanoferrate	Fluorometric
	Trihydroxyindole reaction	Fluorometric

the use of reaction detectors can be summarized in general terms as follows:

1 The derivatizing reagent must be miscible with the mobile phase.
2 The derivatization reaction must be rapid.
3 Provisions must be made for continuous addition of precisely measured volumes of reagents to the column effluent.
4 Adequate mixing of the reaction mixture is required prior to the passage of the mixture through the detector.
5 Provisions must be made for maintaining the temperature at a desired level if incubation must be carried out at a controlled temperature.

The use of reaction detectors minimizes some of the problems associated with the formation of reaction by-products that diminish the utility of precolumn derivatization (38). However, like any other technique, this type of postcolumn derivatization also has some disadvantages. The main limitation lies in the restrictions placed on the composition of the mobile phase, since it must be compatible with the reaction medium. Thus postcolumn reactions cannot be used with many gradient elution programs. In addition, it must be possible to detect the reaction products under different conditions from those of the reactants. Depending on the incubation time necessary to complete the reaction, three types of "reactors" can be used:

1 Short, narrow bore tubing (0.5 mm) (tubular reactors) for incubation times of 10 to 30 sec.
2 Short columns (bed reactors) packed with glass beads (size is similar to the packing material in an analytical column) for incubation times up to 3 min.
3 Long, narrow bore tubing (0.5 mm) with air segmentation of the reaction stream for incubation times up to 20 min.

Examples of applications of reaction detectors are available in the literature (39).

In order to minimize band broadening in tubular reactors with nonsegmented flow, the use of coiled capillaries has been suggested (32). These reactors are useful for fast reactions. The bed reactors, which are used for reactions with intermediate times, should be packed with small particles in order to minimize band broadening resulting from axial dispersion and convective mixing. Packing procedures similar to those used for HPLC columns (32,33) are usually employed.

The air-segmented reactors are becoming increasingly popular for relatively slow reactions (reaction time >5 min.) (34–37). The column effluent is intercepted at certain time intervals by the introduction of air bubbles. This minimizes axial diffusion and thus reduces band broadening. However, transfer of solute molecules from one segment to another cannot be completely avoided.

Reaction detectors can utilize many detection principles: UV absorption, fluorescence, chemiluminescence, electrochemical and photochemical reactions, and so on. If properly designed, they cause minimal band broadening. An added advantage of these detectors is that complete reactions are not necessary as long as the reactions are reproducible. In selecting a detector for a specific reaction, reaction kinetics, pressure and temperature requirements, and corrosiveness of chemicals must be considered. Since these detectors can be used in a variety of applications, they will find increasing use in the future, particularly in the analysis of complex mixtures of trace compounds, where enhanced detectability is required.

REFERENCES

1. Sir. G. Taylor, *Proc. Roy. Soc.*, **A219**, 186 (1953).
2. J. H. Knox and A. Pryde, *J. Chromatogr.*, **112**, 171 (1975).
3. J. H. Knox, *J. Chromatogr. Sci.*, **15**, 352 (1977).
4. J. N. Done, *J. Chromatogr.*, **125**, 43 (1976).
5. R. P. W. Scott and P. Kucera, *J. Chromatogr. Sci.*, **9**, 641 (1971).
6. J. C. Sternberg, in *Advances in Chromatography*, Vol. 2, Dekker, New York, 1966, p. 206.
7. J. C. Kraak, in *Instrumentation for High-Performance Liquid Chromatography*, J. F. K. Huber, Ed., Elsevier, New York, 1978.
8. B. L. Karger, D. P. Kirby, P. Vouros, R. L. Foltz, and B. Hidy, *Anal. Chem.*, **51**(14), 2325 (1979).
9. P. J. Arpino and G. Guiochon, *Anal. Chem.*, **51**(7), 683A (1979).
10. P. J. Arpino, H. Colin, and G. Guiochon, "On-line Liquid Chromatography-Mass Spectrometry," Paper A8, 25th Annual Conference on Mass Spectrometry, Washington, D.C., May 1977.

11. C. R. Blakeley, M. J. McAdams, and M. L. Vestal, *J. Chromatogr.*, **158**, 261 (1978).

12. D. Henneberg, U. Henrichs, H. Husmann, and G. Schomburg, *J. Chromatogr.*, **167**, 139 (1978).

13. D. I. Carroll, I. Dzidic, R. N. Stillwell, K. D. Haegele, and E. C. Horning, *Anal. Chem.*, **47**, 2369 (1975).

14. F. W. McLafferty, R. Knutti, R. Venkataraghavan, P. J. Arpino, and B. G. Dawkins, *Anal. Chem.*, **47**, 1503 (1975).

15. W. H. McFadden, H. L. Schwartz, and S. Evans, *J. Chromatogr.*, **122**, 389 (1976).

16. W. L. Erdahl and O. S. Privett, *Lipids*, **12**, 797 (1977).

17. C. R. Blakley and M. L. Vestal, "Application of Crossed-Beam LC/MS to Involatile Biological Samples," Paper D7, 25th Annual Conference on Mass Spectrometry, Washington, D.C., May 1977.

18. R. P. W. Scott, C. G. Scott, M. Munroe, and J. Hess, *J. Chromatogr.*, **99**, 395 (1974).

19. W. H. McFadden, D. C. Bradford, D. E. Games, and J. L. Gauer, *Am. Lab.*, **9**, 55 (1977).

20. R. E. Lovins, J. Craig, F. Thomas, and C. H. McKinney, *Anal. Biochem.*, **47**, 539 (1972).

21. P. R. Jones and S. K. Yang, *Anal. Chem.*, **47**, 1000 (1975).

22. N. A. Parris, *Instrumental Liquid Chromatography*, Elsevier, New York, 1976, Chap. 4.

23. M. Martin and G. Guiochon, in *Instrumentation for High-Performance Liquid Chromatography*, J. F. K. Huber, Ed., Elsevier, New York, 1979.

24. J. C. Kraak, in *Instrumentation for High-Performance Liquid Chromatography*, J. F. K. Huber, Ed., Elsevier, New York, 1978.

25. E. Kenndler and E. R. Schmid, in *Instrumentation for High-Performance Liquid Chromatography*, J. F. K. Huber, Ed., Elsevier, New York, 1978.

26. C. F. Simpson, Ed., *Practical High Performance Liquid Chromatography*, Heyden, London, 1976.

27. J. H. Knox, J. N. Done, and J. Loheac, *Applications of High Speed Liquid Chromatography*, Wiley, London, 1975.

28. R. P. W. Scott, *Contemporary Liquid Chromatography*, Wiley-Interscience, New York, 1976.

29. R. P. W. Scott, *Liquid Chromatography Detectors*, Elsevier, New York, 1977.

30. R. J. Hamilton and P. A. Sewell, *Introduction to High Performance Liquid Chromatography*, Chapman and Hall, London (Wiley), 1978.

31. J. L. Lawrence, *J. Chromatogr. Sci.*, **17**, 147 (1879), and references contained therein.

32. R. S. Deelder, M. G. F. Kroll, A. J. B. Beeren, and J. H. M. van den Berg, *J. Chromatogr.*, **149**, 669 (1978).

33. R. S. Deelder, M. G. F. Kroll, and J. H. M. van den Berg, *J. Chromatogr.*, **125,** 307 (1976).

34. L. J. Skeggs, *Am. J. Clin. Pathol.*, **28,** 311 (1957).

35. L. R. Snyder and H. J. Adler, *Anal. Chem.*, **48,** 1017, 1022 (1976).

36. L. R. Snyder, *J. Chromatogr.*, **125,** 287 (1976).

37. R. W. Deelder and P. J. H. Hendricks, *J. Chromatogr.*, **83,** 343 (1973).

38. R. W. Frei and A. H. M. T. Scholten, *J. Chromatogr. Sci.*, **17,** 152 (1979).

39. H. Poppe, in *Instrumentation for High-Performance Liquid Chromatography*, J. F. K. Huber, Ed., Elsevier, New York, 1978.

BIBLIOGRAPHY

Baumann, F., A. C. Brown III, S. P. Cram, C. H. Hartmann, and J. L. Hendrickson, *J. Chromatogr. Sci.*, **14,** 177 (1977).

Berry, L. and B. L. Karger, *Anal. Chem.*, **45**(9), 819A (1973).

Byrne, S. H., Jr., in *Modern Practice of Liquid Chromatography*, J. J. Kirkland, Ed., Wiley-Interscience, New York, 1971, Chap. 3.

Halász, I., and P. Vogtel, *J. Chromatogr.*, **142,** 241 (1977).

Huber, J. F. K., Ed., *Instrumentation for High-Performance Liquid Chromatography*, *Elsevier, Amsterdam, 1978.*

Karasek, F. W., *Res. Dev.*, 34 (March 1975).

Kissinger, P. T., *Anal., Chem.*, **49,** 447A (1977).

Kissinger, P. T., L. J. Felice, D. J. Miner, C. R. Preddy, and R. E. Shoup, in *Contemporary Topics in Analytical and Clinical Chemistry*, Vol. 2, D. M. Hercules et al., Eds., Plenum, New York, 1978, p. 55.

Maggs, R. J., in *Practical High Performance Liquid Chromatography*, C. F. Simpson, Ed., Heyden, London, 1976, p. 269.

Martin, M., C. Eon, and G. Guiochon, *J. Chromatogr.*, **108,** 229 (1975).

McNair, H. M. and C. D. Chandler, *J. Chromatogr. Sci.*, **14,** 477 (1977).

Parris, N. A., *Instrumental Liquid Chromatography*, Elsevier, New York, 1976.

Pungor, E., K. Toth, Z. Feher, G. Nagy, and M. Varadi, *Anal. Lett.*, **8,** IX (1975).

Scott, R. P. W., *Liquid Chromatography Detectors*, Elsevier, New York, 1977.

Snyder, L. L. and J. J. Kirkland, *Introduction to Modern Liquid Chromatography*, 2nd ed., Wiley-Interscience, New York, 1979, Chap. 3.

Swartzfager, D. G., *Anal. Chem.*, **48,** 2189 (1976).

IV Columns and Column Performance

The use of nonpolar stationary phases and polar mobile phases in chromatographic systems was first suggested by Boscott in 1947 (1).* A year later, Boldingh (2) used a column of rubber powder and aqueous methanol and water to separate long-chain fatty acids. The technique was named *reversed phase* by Howard and Martin (3), who also used *n*-octane and liquid paraffin stationary phases to separate fatty acids. Reversed phase (RP) was used mainly for separations of nonpolar solutes until Martin and Porter (4) used this technique to fractionate ribonuclease. Until the mid-1960s, the utility of both normal and RP chromatography was plagued by column instability. While mechanically coated stationary phases were rarely subject to significant bleeding in GC, in LC the problems were much more serious. Because of the nature of the mobile phases commonly employed in LC, stationary phases were subject to dissolution in solvents. In addition, because of the weak forces at the interface of the mechanically held stationary phase and the support material, the use of high-solvent flow rates caused excessive bleeding or stripping of the stationary phase, due to the large shear forces. Furthermore, the heat generated enhanced the miscibility of the phases. The use of such systems also restricted the useful range of capacity factors, as well as the use of gradient elution and elevated temperatures.

These problems were solved by the development of bonded phases. Chemical attachment of stationary phases to supports was first described for use in GC (5), and it was suggested (6) that similar packings could greatly augment the versatility and lifetime of LC columns. The ascent of RPLC began with the work of Halász and Sebastian (7), who popularized this method of chemical bonding of stationary phases. This major breakthrough in column technology was further enhanced by the development of pellicular materials, first by Horváth, Preiss, and Lipsky (8)

*References for Sections 1–5 of Chapter IV are listed on pages 106–109.

and later by Kirkland (9), and the introduction of microparticulate packings, popularized by Majors (10) and Hartwick and Brown (11).

Pellicular packings (30- to 40-μm range), which consist of a solid core
and a thin, porous outer shell (retention layer or stationary phase), were
noted for their rapid solute mass transfer, good efficiency, and relative
ease with which they could be packed. However, because of their small
surface areas, they had limited sample capacity (Fig. 34a).

Totally porous, chemically bonded microparticular packings (5- to 10-
μm range) offer significant improvements in column efficiency. The improvements are due to the increased rate of mass transfer which results
from small interparticle spaces and short pore depths (Fig. 34b and c).
More important, because of the increased surface area, their loading
capacity is greater than the pellicular packings. Detailed reviews of HPLC
packings and columns are available in the literature (see Bibliography).
Thus chemically bonded microparticular packings led to the rapid ascent
and tremendous popularity of the RP mode of HPLC (RPLC). The advantages of RPLC are summarized in Table 7.

The great operational simplicity of RPLC systems stems from the weak
surface energies of the bonded phase; hence, the attractive forces between
the solute molecules, stationary phase, and the mobile phase are weak
(12). The analyses are rapid and reequilibration times during solvent
changeover are short. Usually, only 5 to 10 column volumes are necessary
for reequilibration, which is considerably less than in normal phase chromatography. This is particularly advantageous during the process of
method development and routine analyses.

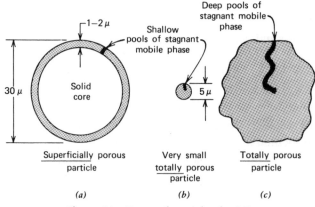

Figure 34   Types of particles for LC.

Table 7 Advantages of RPLC

1. Operational simplicity.

2. Availability of columns with high efficiency and selectivity.

3. Wide range of solvent systems.

4. Compatability of solvent systems with many samples, particularly those containing biological molecules.

5. Rapid analysis and fast equilibration.

6. Possible use of secondary chemical equilibria (ionization, ion-pair and liquid exchange).

7. Compatability with detection devices necessitating aqueous media (electro-chemical, post-column derivatization).

8. Large number of commercially available reversed-phase packings.

9. The use of reversed-phase as a tool for physio-chemical measurements (ionization constants, complex formation constants, hydrophobicity and purity determinations, study of structure-activity relationships for pharmaceutical compounds and antimetabolites).

Since the RP systems utilize both the hydrophobic and polar solute characteristics, it is possible to analyze a larger number of substances of a wide range of polarity and molecular weight. This makes RPLC particularly attractive in biochemistry and clinical chemistry. If the initial mobile phase is a weak (polar) solvent, large volumes of aqueous samples containing low levels of compounds of low polarity can be injected directly into RPLC columns and preconcentrated at the head of the RP column. Next, the solvent strength is sharply increased by means of gradient elution or step-gradient change. This technique, called *trace enrichment,* enhances the detection sensitivity of trace compounds. Sometimes it is also possible to inject untreated biological samples (serum, urine) onto a RP column directly or through a short precolumn. Since buffered water is usually employed as a mobile phase in RPLC, injection of aqueous solutions will not affect retention characteristics. In addition, irreversible adsorption of solutes is rare and highly polar or ionic substances elute in the void volume, thereby reducing possible interferences.

It is becoming increasingly apparent that RPLC can be an important tool for obtaining physicochemical parameters (13–16). Free energies of interactions between the solute and the two phases in a chromatographic system can be deduced from the retention behavior and used for predicting the molecular nature of complex pharmaceuticals and antimetabolites (17). In addition, fundamental molecular properties, such as dissociation constants, complex formation constants (16), and hydrophobic character (14), of many compounds can be determined using RPLC data.

RP columns also have some limitations. Among them, the pH range over which RP columns have long-term stability is between ~2 and 7.5. Eluents of extremely low pH values attack the Si—C bonds, whereas extended use of highly basic solvents (pH >8) leads to the dissolution of the silica matrix. However, the pH restriction does not, in fact, diminish the success of RPLC in separating compounds that differ widely in polarities, since most separations can be achieved within this range if secondary equilibria are employed.

Another problem results from incomplete coverage of the silica surface and the presence of surface silanol groups (Si—OH). Thus the retention of polar compounds, such as protonated amines, proceeds via adsorption and/or ion exchange. These mixed mechanisms usually lead to peak tailing due to slow mass transfer and nonlinearity of the sorption isotherm. The presence of unreacted silanol groups also accelerates the dissolution of silica, which can be hydrolyzed by water according to the following reaction:

$$(OH)_3Si\text{—}O\text{—}Si(OH)_3 + H_2O \rightleftarrows Si(OH)_4 + Si(OH)_4 \qquad (70)$$

To improve column stability and achieve reproducible retention and selectivity, unreacted silanol groups can be treated with a small silating reagent, such as trimethylchlorosilane (TMCS), according to the reaction:

$$\underset{\displaystyle CH_3}{\overset{\displaystyle CH_3}{>\!Si\text{—}OH + Cl\text{—}\underset{|}{\overset{|}{Si}}\text{—}CH_3 \rightarrow\ >\!Si\text{—}O\text{—}\underset{|}{\overset{|}{Si}}\text{—}CH_3 + HCl}} \qquad (71)$$

The influence of residual silanol groups is illustrated with the retention behavior of some PTH amino acids, before and after silanization of an ODS column, as is shown in Table 8.

Table 8   Retention of PTH-Amino Acids[a]

Column: ULTRASPHERE (150×4.6mm)
Mobile phase: 50:50 CH_3OH/H_2O, 0.01m sodium acetate, pH 4.5
Flow rate: 1.0 ml/min
Temperature: Ambient

Amino acid	R group	k' before silanization	after silanization
Valine	$-CH{<}^{CH_3}_{CH_3}$	4.34	4.73
Arginine	$-CH_2-(CH_2)_2-NH-\underset{\underset{NH}{\|\|}}{C}-NH_2$	4.33	1.67
Histidine	$-CH_2-\underset{\underset{\underset{\underset{H}{C}}{\backslash\!/}}{NHN}}{C}{=}CH$	2.49	1.08

[a]Reproduced from reference 62 with permission.

In spite of the great strides made in column technology, there is still a need for columns with better column-to-column and manufacturer-to-manufacturer reproducibility in terms of retention, efficiency, and selectivity. Variations in column performance and the relatively high price of most RPLC columns have aroused considerable customer interest in packing procedures, and many laboratories now routinely pack their own columns.

1   SUPPORTS

Until 1969, the development of RPLC was hampered by the lack of suitable supports. Today, several materials are available for these purposes: silica gel, graphitized and nongraphitized modified carbon blacks, and carbon-coated silica. At the present time, silica is the most popular support because it has many advantages. The important properties of chromatographic silica have been described by several authors (18–22).

1.1 Silica

Silica gel is a polymer of silicic acid and has a general formula $SiO_2 \times H_2O$. It is usually produced by acidic hydrolysis of sodium silicate, followed by polycondensation and dehydration of orthosilicic acid. The primary particles produced condense with one another forming a hydrogel. Since the size of the primary particles determines the surface area and the porosity of the final product, it is important to control the particle size during acid gelation by means of buffer pH and the reaction temperature.

Upon heating the hydrogel at approximately 120°C, further condensation and dehydration occur and the product is known as xerogel. A network of SiO_4 tetrahedra is formed with most hydroxyl groups remaining on the surface. If this material is heated at approximately 150°C for a given period of time, physically sorbed surface water is removed (24). Removal of structural water takes place between 150 and 600°C, and at 600°C, recrystallization occurs. Between 200 and 500°C, free silanol groups condense, leaving mostly the siloxane (Si—O—Si) groups exposed at the surface (18):

$$2 \geqslant\!\!Si—OH \underset{-H_2O}{\rightleftharpoons} \geqslant\!\!Si—O—Si\!\!\leqslant \qquad (72)$$

This reaction is reversible, and siloxane groups react readily with water and alcohols. Differences in some properties of various commercial silicas are summarized in Table 9 and discussed in great detail in the work of Scott and Kucera (24) and Unger (19,25,30).

Fully hydroxylated silica, obtained in the temperature range between 200 and 400°C, is the desirable form for most chromatographic purposes. According to several authors (26–28), there are five different types of sites on silica:

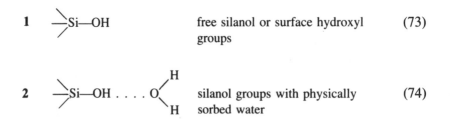

| 1 | $\diagdown\!\!\!\!\diagup\!Si—OH$ | free silanol or surface hydroxyl groups | (73) |

| 2 | $\diagdown\!\!\!\!\diagup\!Si—OH \ldots O\langle^H_H$ | silanol groups with physically sorbed water | (74) |

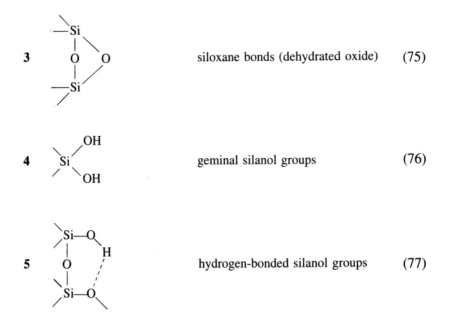

3   siloxane bonds (dehydrated oxide)   (75)

4   geminal silanol groups   (76)

5   hydrogen-bonded silanol groups   (77)

Table 9   Properties of Commercially Available Silica[a]

Type	Name	Surface Area (m²/g)	Particle Size (μm)	Shape*	Supplier
Silica Pellicular	Corasil II	14	37–50	S	Waters
	Vydac adsorbent**	12	30–44	S	Separations Group
	Pellosil HS	4	37–44	S	Whatman
	Pellosil HC	8	37–44	S	Whatman
	Perisorb A	14	30–40	S	Merck***
	SIL-X-II**	12	30–40	S	Perkin-Elmer
Porous	Porasil	400	10	I	Waters
	Silica A	400	13 ± 5	I	Perkin-Elmer
	SIL-X-I**	400	13 ± 5	I	Perkin-Elmer
	LiChrosorb@ SI-60	500	5 or 10	I	Merck
	LiChrosorb@ SI-100	400	5 or 10	I	Merck
	Spherisorb S5W, S10W, S20W	200	5, 10 or 20	S	Phase Separation
	Partisil 5, 10, 20	400	5, 10 or 20	I	Whatman
	Zorbax SIL	300	4–6	S	DuPont
	Porasil T	300	15–25	I	Waters
	LiChrospher SI-100	250	10	S	Merck
	LiChrospher SI-500	50	10	S	Merck
	LiChrospher SI-1000	20	10	S	Merck
	LiChrospher SI-4000	6	10	S	Merck
	Micropak SI-5, SI-10	500	5 or 10	I	Varian

*I = Irregular; S = spherical
**Stated to be chemically deactivated; control of water content in system is less critical.
***E.M. Labs in the U.S.A.
@Formally marketed under the name Merckosorb.

[a]Reproduced (with modification) from reference 78 with permission.

The silanol groups are weakly acidic ($pK_a \simeq 9$), which means that silica can act as a weak cation exchanger:

$$\equiv Si-OH \rightleftharpoons \equiv Si-O^- + H^+ \tag{78}$$

There are, however, slight variations in the acidity of various groups, and the most acidic silanol groups are the ones with intramolecular hydrogen bonding. Most of the chromatographic properties of the silica surface are derived from the presence of surface silanol groups, although siloxane groups may also contribute to the surface activity (30). The maximum concentration of silanol groups on porous silica, activated at 200°C, is 8.0 to 9.0 $\mu mol/m^{-2}$ (29). The concentration of silanol groups can be determined by reaction with methyllithium and subsequent quantitative determination of methane (29a) or by heterogeneous isotope exchange with D_2O, followed by mass-spectrophotometric analysis (29b). Additional information concerning the presence of free silanol groups can be deduced from IR spectroscopy (sharp band at 3750 cm^{-1}) (30,49).

Different methods for activation of silica (completely hydroxylated) have been described in the literature: acid hydrolysis or hydration in 0.1 M HCl at 90°C for 24 hours, heating at 250°C under vacuum (30), treatment with a hot solution of $SiCl_4$ in purified dioxane (32), or heating with a mixture of nitric acid and sulfuric acid at 100°C for 12 hours (33). The surface areas of commercial silica range from 100 to 860 m^2/g, and the average micropore diameter ranges from 35 to 330 Å.

The shape of commercially available silica particles can be either irregular or spherical. The spherical particles have a narrower size distribution and they are believed to give more uniform packings (34). Columns packed with irregularly shaped silica are subject to non-uniform stress distribution, which might lead to gradual dissolution of the bonded organic moieties on sharp edges (13). The relative merits of the two types of silica have not yet been fully investigated. However, reports in the literature indicate that the efficiencies of columns packed with irregularly shaped and spherical silica of equivalent particle diameter are similar (34).

2   TYPES OF BONDED PHASES

Modified adsorbents for RP chromatography are prepared by chemically bonding organic groups to the silica surface. The idea was pioneered by

Halász and co-workers (5), Kirkland and DeStefano (36), Locke and co-authors (37), and others (see Bibliography). According to their structure, chemically bonded phases can be classified as (a) *monomeric* and (b) *polymeric phases*.

The monomeric type, which is also called the "brush type" or "bristle structure," involves the formation of a monomolecular organic layer on the surface of the silica particles; thus a single organic molecule is attached to a single surface site. Due to steric hindrance during the grafting process, not all sites react and some silanol groups are left free or nonbounded.

Because of the rapid mass transfer in the stationary phase, the efficiency of monomeric layer packings is much greater than that of the polymeric phases. Therefore, monochlorosilanes are used to minimize polymerization. However, Karch et al. (38) have observed that greater surface coverage of monomeric phases can be achieved if di- or trichlorosilanes are used under anhydrous conditions, provided a large excess of silane is added and the volatile products are removed continuously (30,39,40). It should also be noted that when polyfunctional organic groups are used, reactions have to be carried out in dry solvents (usually toluene), since traces of water can lead to cross-linked polymerization reactions.

Polymeric phases can be prepared by reacting silica with a di- or trichlorosilane in the presence of trace amounts of water, which hydrolyze $\geqslant$Si—Cl groups into silanol groups. This leads to the formation of a layer structure with properties similar to those of a liquid.

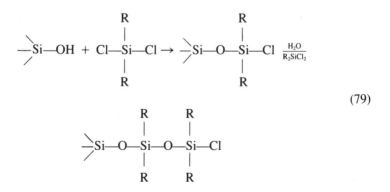

$$(79)$$

Due to the high reactivity of dichloroalkylsilanes, the surface layer is probably a composite of monomeric and dimeric units.

Although packings with polymeric layers are characterized by slow solute mass transfer in the stationary phase and by slow equilibration, they are still used in coating porous layer beads as well as in achieving high coverage of the silica surface (36,41).

Bonded phases can also be classified according to the type of bond between the organic moiety and the silica support. There are essentially four different bonded phases:

1	$\geq Si-O-R$	ester phases
2	$\geq Si-NR_2$	amino phases
3	$\geq Si-CR_3$	carbon phases
4	$\geq Si-O-Si-CR_3$	siloxane phases

The organic (R) group can be either polar or nonpolar. The differences in chromatographic behavior of various commercially available bonded phases result from the following characteristics:

1 Type of organic moiety.
2 Chain length of organic moiety.
3 Amount of organic moiety per unit volume.
4 Particle size and shape of the support.
5 Porosity and surface area of the support.
6 Surface topology of bonded material.
7 Concentration of free silanol groups.

2.1 Ester Phases

Packings of this type, first introduced by Halász and Sebastian (35), are typically prepared by the reaction of surface silanols with alcohols:

$$\geq Si-OH + HOR \underset{-H_2O}{\rightleftharpoons} \geq Si-O-R \qquad (80)$$

These packings are rarely used today because of their thermal instability at elevated temperatures and their tendency to hydrolyze in aqueous solutions.

2.2 Amino Phases

Packings with a bond between the silicon of the support and the nitrogen of an amine were first described in 1950 by Deuel et al. (42). They possess higher thermal and hydrolytic stability than the ester type and can be used with aqueous solvents in the pH range between 3 and 8 (38,43–46).

Prior to the attachment of the amino group, silanol groups are converted into $\geqslant$Si—Cl and then aminated:

$$\geqslant\text{Si—OH} \xrightarrow{SO_2Cl} \geqslant \text{Si—Cl} \xrightarrow[-HCl]{R_2NH} \geqslant\text{Si—NR}_2 \tag{81}$$

The bonded phases thus prepared are monomeric of the brush type. Their properties can be modified by introducing a functional group into the ω position with respect to the amino group. These packings are not commercially available.

2.3 Carbon Phases

The preparation of packing materials of the $\geqslant$Si—CR$_3$ type is usually carried out in a two step reaction: (1) halogenation of the support and (2) reaction with a Grignard reagent (Wurtz or Friedel-Crafts reaction):

1 Halogenation of the support:

$$\geqslant\text{Si—OH} + \text{SOCl}_2 \rightarrow \geqslant\text{Si—Cl} \tag{82}$$

2 Reaction:

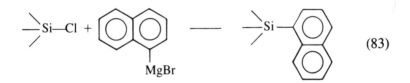

$$\tag{83}$$

Further reaction with Br$_2$, followed by reaction with another Grignard reagent, would yield a polynaphthyl.

Locke et al. (37) give a detailed account of the experimental procedure. Both steps in the reaction can be repeated in order to obtain a larger content of the organic moiety.

Halśz and Sebastian (47) used a different approach in the preparation of chemically bonded phases containing the Si—C bond. They silanized the silica gel support with dimethyldichlorosilane and then subjected hexamethyldisilane to chlorobromination, followed by a reaction with diethylenetriamine. The product had the following structure:

$$\geqq Si—CH_2—NH—(CH_2)_2—NH—(CH_2)_2—NH_2 \qquad (84)$$

With dibutyldichlorosilane and N,N-dimethylethylenediamine, the phase obtained had this structure:

$$\geqq Si—(CH_2)_4—NH—(CH_2)_2—N(CH_3)_2 \qquad (85)$$

These monomeric phases of the brush type are more thermally and hydrolytically stable than the ester phases, and the mass transfer is very rapid. The Si—C bond is unstable in very basic solutions.

2.4 Siloxane Phases

Chemically bonded phases in which the stationary phase is connected with the support through a siloxane (Si—O—Si) bond are usually prepared by the reaction of surface silanol groups with an organochlorosilane:

$$
\begin{array}{ccc}
& R_1 & R_1 \\
& | & | \\
\geqq Si—OH + Cl—Si—R & \rightarrow & \geqq Si—O—Si—R + HCl \qquad (86) \\
& | & | \\
& R_2 & R_2
\end{array}
$$

Reaction with monochlorosilanes is preferred to the ones with di- or trichloro analogs, since it does not give rise to polymerization. Usually the R group is an octadecyl chain [$—(CH_2)_{17}CH_3$], and ODS phases, one of the most popular types of siloxane bond packings, are thus obtained. Other moieties such as amine [$—(CH_2)_nNH_2$], nitrile [$—(CH_2)_nCN$], phenyl [$—(CH_2)_nC_6H_5$], ethyl ($—CH_2—CH_3$), or octyl [$—(CH_2)_7—CH_3$] can also be grafted to the silica surface.

Since only approximately 45% of the surface silanol groups will be bonded, the remaining free silanol groups can be covered by treatment with trimethylchlorosilane, giving a completely hydrophobic surface.

If di- and trisubstituted silanes are used in the derivatization, the following reactions are possible (48):

1 Disubstituted silanes:

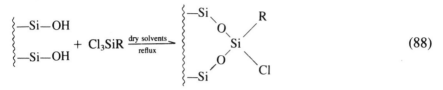

$$(87)$$

2 Trisubstituted silanes in dry solvents:

$$(88)$$

3 Trisubstituted silanes in solvents with protic impurities:

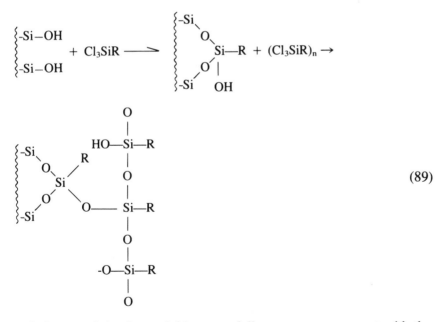

$$(89)$$

Only two of the three trichloro or trialkoxy groups can react with the surface silanol groups, due to steric hindrance (49). The remaining groups can be hydrolyzed to hydroxyl groups if water is not excluded from the solvent system. This in turn leads to further reaction with excess reagent and formation of multilayer phases. Although these phases proved to be useful in GC, their properties are less desirable in LC owing to the slow mass transfer in stagnant pockets of trapped mobile phase organosilanes.

2.5 Styrene-Divinylbenzene Copolymers

Styrene-divinylbenzene copolymers are not commonly used in bonded phase chromatography. They offer certain advantages in terms of good thermal stability over a wide pH range (1–12) and physical durability (50). The major problem with these packings is that variations in the column packing accompany the change from an organic to aqueous solvent. However, in the future, copolymers specially designed to produce specific properties may be useful in RPLC.

2.6 Modified Carbon Black

Using the term "reversed phase" in a broader sense to include not only the separation of nonpolar solutes but also homologous series of solutes differing slightly in molecular weight, thermal carbon blacks modified by benzene pyrolysis (20% pyrocarbon) have potential use. The procedures for the preparation of these phases have been discussed in the literature (51). However, higher efficiency and greater stability are needed before these packings can be used routinely.

3   PREPARATION OF RP CHEMICALLY BONDED PHASES

Preparation of chemically bonded phases can be achieved in two ways: by batch-type process and by *in situ* modification.

Each method will be discussed separately. Properly prepared chemically bonded phases should have the following characteristics:

1 The surface of the silica support should be completely covered; otherwise a mixed retention mechanism will result.
2 The composition of the phase should be defined and reproducible.
3 The phase should have high chemical and thermal stability.

3.1   Batch-Type Process

3.1.1   Surface Modification

Surface modification involves a reaction between completely hydroxy-lated silica and a modifier such as organochlorosilane or organoethoxy-silane.

Since the maximum coverage density is obtained with a maximal number of free silanol groups, it is necessary to activate the support prior to the modification. This is usually achieved by heating the silica at approximately 250°C under high vacuum, at which temperature the silica surface is fully hydroxylated. The surface concentration of free silanol groups will be approximately 8.0 μmol/m^2 (4.8 groups/nm^2 (18,53). If the silica is heated above 300°C, formation of siloxane bonds takes place. After addition of the modifier, the mixture is heated at approximately 20°C below the boiling point of the corresponding silane for 8 hours in dry nitrogen atmosphere (30). The bonded phase is then washed and subjected to vacuum at 200°C. Depending on the reaction conditions, the resulting material can have a chemically bonded monolayer, multilayer, or polymer of the organic modifier. Multilayers are formed by repeated reactions of the monolayer with reactive groups and excess modifier, whereas polymeric layers are formed when a polymerizing agent is added to the silica support (52). Due to the experimental difficulties in controlling the extent of cross-linking and homogeneity of both types of layers, the composition of polymeric phases cannot be easily defined (31).

If a monolayer is to be formed, the following requirements must be fulfilled:

1   The silica surface must be completely hydroxylated.
2   Physically sorbed water must be removed, otherwise the modifier may hydrolyze and condense to give a polymer of physically sorbed organosilica.

In order to achieve maximal coverage, an excess of the modifier must be added to the silica. In addition, the reaction time and temperature have to be carefully chosen and controlled in order to increase the reaction rate, which is governed by the diffusion of the organic modifier to the

Table 10   Variation of Silica Surface Coverage with Silanization Time[a]

Parameter	Reaction time (h)					
	0.8	2.5	5	6.5	18	24
Carbon content (%, w/w)	3.0	3.6	4.2	6.5	10.5 11.3	11
Surface concentration (μmol·m^{-2})	0.36	0.45	0.52	0.84	1.46 1.60	1.54

[a]Reproduced from reference 63 with permission.

active sites on silica. The influence of the silanization time on the carbon content is illustrated with the reaction of silica with a 10% solution (v/v) of octadecyltrichlorosilane in refluxing dry toluene (Table 10). Beyond 18 hours, no significant increase in the carbon content was observed (63).

The effects of temperature on the reaction yield are shown in Table 11. It is obvious that higher reaction temperatures lead to a higher carbon content.

It has been mentioned before that monofunctional ($XSiR_3$), bifunctional (X_2SiR_2), or trifunctional (X_3SiR) modifiers can be used. The stoichiometry of the surface reaction can be studied by IR spectroscopy using transmission and the attenuated total reflection techniques (53,54) or by mass spectrometry (55). In the reaction between the silica and the monofunctional modifier, the F factor (30), which is the ratio of the number of moles of silanol groups reacted to the number of moles of the organic modifier, is usually 1. In the case of bifunctional and trifunctional modifiers, the F value has been reported to be $1 \leq F \leq 2$ (30). With increasing chain length and molecular weight of the modifier, F cannot be 3 due to steric reasons. Thus a larger surface area is needed, since binding is governed by the crowding of the groups at the bonding site. Therefore, a decrease in coverage is observed with increasing chain length, probably due to the blockage of the micropores.

Due to the incomplete reactions of bifunctional and trifunctional modifiers, one or two chlorine groups remain unreacted. Upon hydrolysis, the chlorines are replaced by hydroxyl groups. Since hydroxyl groups impart undesirable properties, they should be derivatized.

The presence of free silanol groups can be detected by using one of the following techniques:

1   IR spectroscopy.
2   Isotopic exchange with tritium-labeled water.

3 Methyl red adsorption.

4 Chromatography.

If the coverage is complete, and if the bonded phase is used with a nonpolar solvent such as dry heptane, then the solvents such as methanol, diethyl ether, and acetone should elute near the void volume and their peaks should not exhibit tailing (30). Also, upon additional treatment with the organic modifier, the carbon content of the packing should not increase (30). Finally, if the concentration of free silanol groups is very small, the capacity factors of different solutes will be less sensitive to the water content of the mobile phase. The detrimental effect of unprotected silanol groups is more pronounced with short-chain bonded phases since the penetration of the solute into the stationary phase is easier. It should also be pointed out that upon attachment of the organic modifier, the structural properties of silica are modified. As can be seen from Table 12, the decrease in the pore diameter, which is theoretically equal to the doubled thickness of the layer, leads to a decrease in the specific surface area (S_{BET}) and specific pore volume (V_p) (26,30,31). For a column coated with octadecyl groups, the decrease in specific surface area and specific pore volume is 54 and 61%, respectively. This decrease increases with increasing chain length (31). The capacity factors for different solutes would therefore be expected to decrease with decreasing specific surface area when going from a C_8 to C_{18} bonded phase. However, since this is not observed, it appears that the increase in chain length overcompensates the decrease in the specific surface area. Furthermore, if the quantity log k'/S_{BET} is plotted against the carbon number of the organic modifier, the relationship is linear, indicating that the entire chain length is available for interactions (30).

Table 11   Variation of Maximum Surface Coverage with Silanization Temperature[a]

Parameter	Solvent for silane		
	Toluene	Xylene	Silane alone
Reaction temperature (°C)	111	138	170
Maximal carbon content obtained (%, w/w)	10	15	23
Reaction time necessary to obtain maximal carbon content (h)	18	15	2

[a]Reproduced from reference 63 with permission.

Table 12   Variation of Pore Structure Parameters of Silica by Means of Surface Modification[a]

Type of packing	Specific surf. area		Specific pore vol.		Most frequent pore diameter, D (nm)
	S_{BET} (m^2/g)	$-\Delta S_{BET}$ (%)	V_p (ml/g)	$-\Delta V_p$ (%)	
Original SiO_2	300	–	1.32	–	12.6
Octylsilyl groups bonded	201	33	0.82	38	11.4
Octadecylsilyl groups bonded	139	54	0.52	61	7.0; 11.4

[a]Reproduced from reference 30 with permission.

3.1.2   Bulk Modification

In the bulk modification procedure, partial hydrolysis and co-condensation of organotriethoxysilane yields a viscous liquid of poly (organoethoxy) silane, which is then emulsified in a water-ethanol mixture. After the addition of a basic catalyst, the hydrolysis and polycondensation are completed, and the hydrogel is dehydrated. The product is composed of porous beads with covalently bonded modifier in the bulk of the particle framework, as well as at the surface. It should be pointed out that the use of organotriethoxysilane alone results in the formation of finely divided powders instead of mechanically stable, totally porous microsilica (30). In addition, at concentrations higher than approximately 30 mol % of organotriethoxysilane, the product exhibits considerable swelling, making it unsuitable as a packing material for LC.

The bulk modification procedure is advantageous, since it permits the control of the pore structure and the surface area of the microbeads. The porosity of the beads is controlled by the amount of solvent added, and the specific surface area depends on the type and concentration of the basic catalysts. Details of the preparation procedure have been described in the literature (31).

The capacity factors of different solutes on the bulk-modified packings are higher than the surface-modified packings. However, there is no difference in selectivity. Retention of bulk-modified materials can be lowered by decreasing the water content of the mobile phase.

3.2   In Situ Preparation of Bonded Phases

The formation of most chemically bonded phases is done using a batch-type process. However, in order to obtain a more easily controlled and consistent bonding, Gilpin et al. (56–58) have proposed a system for a totally *in situ* preparation of monomeric and polymeric phases. The use of dry solvents in the coating procedure is important since prereaction surface moisture was found to be a critical factor in determining the efficiency of the resulting column. The optimal amount of physically adsorbed water is at a point immediately prior to its complete removal. Prior to the *in situ* modification of the slurry-packed silica, the column is flushed first with a dry toluene. Next, a solution of organosilane in dry toluene is pumped through the system. In order to ensure reproducible surface coverage, an excess of chlorosilane is used. Additional dry toluene is then pumped through the column to ensure the removal of unreacted organosilane. Multilayer coatings can be prepared in a similar manner by repeating the silanization step and rinsing the column between successive coatings. Overall column performance is evaluated by calculating the ratio of the HETP value obtained before chemical modification to the HETP value obtained after modification.

4 SURFACE COVERAGE

Since the exact molecular surface topology of bonded phases is not known, the amount of the alkyl silica in the stationary phase is an important parameter that can be used to compare the performance of different columns. This parameter can be expressed in several ways:

1 *Weight-per-weight percent carbon.* Typical values for microparticulate silica with a surface area of 300 to 350 m²/g are 3 to 4% and 20% for methylsilane and octadecylsilane, respectively. This method of characterizing the bonded phase is not appropriate since the percent of carbon will depend on the surface area and whether or not silica was fully hydroxylated prior to bonding (31).

2 *Carbon content per unit specific surface area.* Using the carbon analysis and the specific surface area, coverage can be expressed in terms of α, the concentration of the surface bonded species (30):

$$\alpha_{exp}(mol/m^2) = \frac{W}{MS_{BET}} \tag{94}$$

where W is the weight of functional groups per weight of adsorbent, M the molar weight of the bonded modifier (g/mol), and S_{BET} is

Table 13 Coverage Densities of Bonded Phases[a]

Concentration of free silanols (μmol/m^2)	Silica	Reagent	Coverage density of bonded groups (μmol/m^2)	Reference
8	LiChrosorb	Chlorosilane–CH$_3$	8.7	38
		–n-C$_4$	4.9	
		–n-C$_{10}$	3.8; 3.5	
		–n-C$_{18}$	2.4; 2.5; 2.6; 2.9	
8	Home-made	Chlorosilane–(CH$_3$)$_3$	4.5**; 4.1***	31
		–(CH$_3$)$_2$Ph*	2.6; –	
		–(Ph)$_3$	1.5; –	
		–(CH$_3$)$_2$–nC$_4$	3.6; 3.7	
		–(pH)$_2$–nC$_4$	1.8; 1.7	
		–(CH$_3$)$_2$–nC$_8$	3.8; 3.4	
		–(CH$_3$)$_2$–nC$_{16}$	3.4; 3.0	
	Spherosil XOA 400	Chlorosilane–(CH$_3$)$_3$	2.51	40
		–(CH$_3$)$_2$Ph	2.5	
		Trichlorosilane–n-C$_{18}$	3.13	
	PSM 50	Chlorosilane–(CH$_3$)$_3$	2.49	
		–(CH$_3$)$_2$Ph	2.36	
		Trichlorosilane–n-C$_{18}$	3.10	
	PSM 500	Chlorosilane–(CH$_3$)$_3$	3.25	
		–(CH$_3$)$_2$Ph	3.08	
		Trichlorosilane–n-C$_{18}$	4.00	
8.3; 10	Porasil A→F	TMCS	3.85; 6.64 (depending on drying temp.)	27
		DMCS	3.47; 3.58 (depending on specific surface area)	

LiChrosorb SI-100	Trichlorosilane	
	-Ph	3.78
	-$(CH_2)_2$Ph	3.70
	-$(CH_2)_4$Ph	3.78
	-$(CH_2)_6$Ph	4.13
	-n-C_8	3.74
	-n-C_{11}	3.42
	-n-C_{13}	3.59
	-n-C_{15}	3.47
	-n-C_{18}	3.44
	-n-C_{21}	3.35

*Ph = phenyl.
**Silica of 211 m^2/g.
***Silica of 301 m^2/g.
aReproduced from reference 60 with permission of Preston Publications Inc.

the specific surface area of silica (m^2/g) starting prior to the modification. However, uncertainties can arise with this procedure since the N_2 sorption method used to measure the surface area usually includes micropores that may not all be accessible to the silanizing agent.

Locke (22) proposed yet another way of expressing the surface coverage. The extent of the reaction is reported in terms of the percent of surface silanol groups reacted with the modifier. The information can be obtained from measurements of the free silanol groups (by IR spectroscopy) before and after the reaction (60). This method is based on the following assumptions:

1   Approximately four free silanol groups per 100 Å are present.
2   Only 70 to 80% of them will react with the organic modifier.
3   One molecule of the tri- and bifunctional modifier reacts with two surface silanol groups, and only monolayers are formed.

Literature values of coverage densities for more bonded phases are reported in Table 13.

REFERENCES

1.   R. J. Boscott, *Nature* (London), **159,** 342 (1947).
2.   J. Boldingh, *Experientia,* **4,** 270 (1948).
3.   G. A. Howard and A. J. P. Martin, *Biochem. J.,* **46,** 532 (1950).
4.   A. J. P. Martin and R. R. Porter, *Biochem. J.,* **49,** 215 (1951).
5.   E. W. Abel, F. H. Pollard, P. C. Uden, and G. Nickless, *J. Chromatogr.,* **22,** 27 (1966).
6.   H. N. M. Stewart and S. G. Perry, *J. Chromatogr.,* **37,** 97 (1968).
7.   I. Halász and I. Sebastian, *Angew. Chem.,* **8,** 453 (1969).
8.   C. Horváth, B. Preiss, and S. R. Lipsky, *Anal. Chem.,* **39,** 1422 (1967).
9.   J. J. Kirkland, *J. Chromatogr. Sci.,* **7,** 7 (1969).
10.   R. E. Majors, *Anal. Chem.,* **44,** 1722 (1972).
11.   R. A. Hartwick and P. R. Brown, *J. Chromatogr.,* **121,** 251 (1976).
12.   B. L. Karger and R. W. Giese, *Anal. Chem.,* **50**(12), 1048A (1978).
13.   C. Horváth and W. Melander, *J. Chromatogr. Sci.,* **15,** 393 (1977).
14.   E. Tomlinson, *J. Chromatogr.,* **113,** 1 (1975).

15. N. Tanaka and E. R. Thornton, *J. Am. Chem. Soc.*, **99**, 7300 (1977).
16. C. Horváth, W. Melander, and A. Nahum, in *Advances in Chromatography*, A. Zlatkis, Ed., Proc. of the 14th Int. Symp., Lausanne, Sept. 24–28, 1979.
17. C. Horváth, W. Melander, and J. Molnár, "Hetaeric Liquid Chromatography," Abstract, 031, 173rd Natl. Meeting, ACS, New Orleans, March 20–25, 1977.
18. K. Unger, *Angew. Chem. Int. Edit.*, **11**(4), 267 (1972).
19. K. Unger, *Porous Silica: Its Properties and Use as Support in Column Liquid Chromatography*, Elsevier, New York, 1979.
20. L. R. Snyder, *Principles of Adsorption Chromatography*, Dekker, New York, 1968.
21. K. K. Unger, W. Messer, and K. F. Krebs, *J. Chromatogr.*, **149**, 1 (1978).
22. D. C. Locke, *J. Chromatogr. Sci.*, **11**, 120 (1973).
23. W. K. Lowen and E. C. Broge, *J. Phys. Chem.*, **65**, 16 (1961).
24. R. P. W. Scott and P. Kucera, *J. Chromatogr. Sci.*, **12**, 473 (1974); *B*, 337 (1975).
25. K. K. Unger, W. Messer, and K. F. Krebs, *J. Chromatogr.*, **149**, 1 (1978).
26. R. K. Gilpin and M. F. Burke, *Anal. Chem.*, **45**, 1383 (1973).
27. A. Diez-Cascon, A. Serra, J. Pascual, and M. Gassiot, *J. Chromatogr. Sci.*, **12**, 559 (1974).
28. L. R. Snyder and J. W. Ward, *J. Phys. Chem.*, **70**, 3941 (1966).
29. L. T. Zhuravlev and A. V. Kiselev, *Zh. Fiz. Khim.*, **39**, 236 (1965).
29a. K. Unger and E. Gallei, Kolloid-Z., *Z. Polymere*, **237**, 350 (1970).
29b. L. T. Zhuravelev and A. V. Kiselev, *Kolloidnyi Zh.*, **24**, 22 (1962).
30. K. K. Unger, N. Becker, and P. Roumeliotis, *J. Chromatogr.*, **125**, 115 (1976).
31. G. Erdel, K. Unger, H. Fischer, and B. Straube, in *Proceedings of the International Symposium, RILEM/IUPAC, on Pore Structure and Properties of Materials*, Vol. III, B-127, Academia, S. Modry and M. Svata, Eds., Prague, 1974.
32. R. S. Deedler, W. M. Claassen, and P. J. H. Hendricks, *J. Chromatogr.*, **91**, 201 (1974).
33. D. G. I. Kingston and B. B. Gerhart, *J. Chromatogr.*, **116**, 182 (1976).
34. G. L. Laird, J. Jurand and J. H. Knox, *Proc. Soc. Anal. Chem.*, **12**, 311 (1974).
35. I. Halász and I. Sebastian, *Angew. Chem. Int. Edit.*, **8**, 653 (1969).
36. J. J. Kirkland and J. J. DeStefano, *J. Chromatogr. Sci.*, **8**, 309 (1970).
37. D. C. Locke, J. J. Schermud, and B. Banner, *Anal. Chem.*, **44**, 90 (1972).
38. K. Karch, I. Sebastian, and I. Halász, *J. Chromatogr.*, **122**, 3 (1976).
39. J. J. Kirkland and P. C. Yates, U.S. Patent 3,722, 181, March 27, 1973; U.S. Patent 3,795,313, March 5, 1974.
40. J. J. Kirkland, *Chromatographia*, **8**, 661 (1975).
41. R. E. Majors, *Analusis*, **10**, 549 (1975).
42. H. Deuel, G. Huber, and R. Iberg, *Helv. Chim. Acta*, **33**, 1229 (1950).
43. A. Pryde, *J. Chromatogr. Sci.*, **12**, 486 (1974).
44. H. Colin, C. Eon, and G. Guiochon, *J. Chromatogr.*, **119**, 41 (1976); **122**, 223 (1976).

45. H. Colin and G. Guiochon, *J. Chromatogr.*, **126**, 43 (1976); **137**, 19 (1977).

46. A. Nakae and G. Muto, *J. Chromatogr.*, **120**, 47 (1976).

47. I. Halász and I. Sebastian, *Chromatographia*, **7**, 371 (1974).

48. E. Grushka and E. J. Kikta, Jr., *Anal. Chem.*, **49**(12), 1005A (1977).

49. H. Hemetsberger, W. Maasfeld, and H. Ricken, *Chromatographia*, **7**, 303 (1976).

50. M. D. Grieser and D. J. Pietrzyk, *Anal. Chem.*, **45**, 1383 (1973).

51. H. Colin, C. Eon, and G. Guiochon, *J. Chromatogr.*, **122**, 223 (1976).

52. E. Grushka, Ed., *Bonded Stationary Phases in Chromatography*, Ann Arbor Science Publishers Inc., Ann Arbor, MI, 1974.

53. A. V. Kiselev and V. I. Lygin, *Infrared Spectra of Surface Compounds*, Wiley-Interscience, New York, 1975.

54. A. Ahmed, E. Gallei, and K. Unger, *Ber. Bunsenges Phys. Chem.*, **79**, 66 (1975).

55. L. T. Zhuravlev, V. A. Kiselev, and V. P. Naidina, *Russ. J. Phys. Chem.*, **42**, 1200 (1968).

56. R. K. Gilpin, J. A. Korpi, and C. A. Janicki, *Anal. Chem.*, **46**, 1314 (1974).

57. R. K. Gilpin, D. J. Camillo, and C. A. Janicki, *J. Chromatogr.*, **121**, 13 (1976).

58. K. Unger, N. Becker, and E. Kramer, *Chromatographia*, **8**, 283 (1975).

59. S. Brunauer, P. Emmett, and E. Teller, *J. Am. Chem. Soc.*, **60**, 303 (1938).

60. R. E. Majors and M. J. Hopper, *J. Chromatogr. Sci.*, **12**, 767 (1974).

61. H. Colin and G. Guiochon, *J. Chromatogr.*, **141**, 289 (1977).

62. N. H. C. Cooke and K. Olsen, *Am. Lab.*, **8**, 45 (1979).

63. M. C. Hennion, C. Picard, and M. Caude, *J. Chromatogr.*, **166**, 21 (1978).

64. J. J. Kirkland, *J. Chromatogr. Sci.*, **9**, 206 (1971).

65. W. Strubert, *Chromatographia*, **6**, 50 (1974).

66. R. M. Cassidy, D. S. LeGay, and R. W. Frei, *Anal. Chem.*, **46**, 340 (1974).

67. T. N. Webber and E. H. McKerrell, *J. Chromatogr.*, **122**, 243 (1976).

68. B. Coq, C. Gonnet, and J. L. Rocca, *J. Chromatogr.*, **108**, 249 (1975).

69. J. J. Kirkland and P. E. Antle, *J. Chromatogr., Sci.*, **15**, 137 (1977).

70. P. A. Bristow, *J. Chromatogr.*, **131**, 57 (1977).

71. G. B. Cox, C. R. Liscombe, M. J. Slucott, K. Sudgen, and J. A. Upfield, *J. Chromatogr.*, **117**, 269 (1976).

72. H. R. Linder, H. P. Keller, and R. W. Frei, *J. Chromatogr. Sci.*, **14**, 234 (1976).

73. C. J. Little, A. P. Dale, D. A. Ord, and T. R. Marten, *Anal. Chem.*, **49**, 1311 (1977).

74. S. H. Chang, K. M. Gooding, and F. E. Regnier, *J. Chromatogr.*, **125**, 103 (1976).

75. J. H. Knox, *Intensive Course on High Performance Liquid Chromatography*, Wolfson Liquid Chromatography Unit, Dept. of Chemistry, University of Edinburgh, 1977, p. 163.

76. J. J. Kirkland, *J. Chromatogr. Sci.*, **10**, 593 (1972).

77. R. Endele, I. Halász, and K. Unger, *J. Chromatogr.*, **99**, 377 (1974).

78. N. A. Parris, *Instrumental Liquid Chromatography*, Elsevier, New York, 1976, Chap. 7.

5 COLUMN PACKING PROCEDURES FOR MICROPARTICULATE PACKINGS

At the time when microparticulate packings were introduced, successful packing of columns was considered somewhat of an "art," and only prepacked and pretested columns of guaranteed performance were used. While the factory-packed columns still have widespread use, many laboratories are now packing their own columns. This was made possible by improvements in the packing technology as well as by disclosure of the details of the packing procedures, many of which have been reported in the literature (1–13).

Chemically bonded packing materials of different polarities can be purchased in several particle sizes; the most common particle sizes are 5- and 10-μm diameters. Particle sizes of less than 5 μm result in decreased column permeability which necessitates the use of excessive packing pressures. However, this can be overcome by using shorter columns (5 to 10 cm in length). Commercially available packings for HPLC are usually supplied with information about the particle size distribution. Most often, three points such as 10, 50, and 90% or a normal distribution curve are quoted, indicating the range over which particle sizes exist.

It should be pointed out that for efficient separations and long column lifetime, the distribution curve should be as narrow as possible. Very small particles, some of which may result from attrition of packing during transportation, should be separated by elutriation procedures. Most often, the packing material is suspended in a solvent (ethanol), thoroughly mixed, and allowed to settle. The supernatant solution contains the smallest particles, which, if not removed, would clog the column and give rise to excessive operating procedures.

Since Knox and Parcher (30) and DeStafano and Beachell (14) demonstrated that the wall effects can seriously affect the efficiency of HPLC columns, the use of 1/8-in. O.D., 2.1-mm I.D. stainless steel columns was discontinued. Until recently, modern RPLC columns were usually produced from 1/4-in. O.D., 4.6-mm I.D. stainless steel tubing of lengths up to 30 cm. Longer columns require a larger bore size if the wall effects are to be minimized.

For successful preparation of a column, the steel tubing must be cut so that the ends are perfectly square. The interior of the tubing has to be polished and must be completely free of manufacturing oil and particulate matter. Degreasing is accomplished by rinsing with solvents of increasing polarity; for example, isooctane, toluene, chloroform, acetone, methanol, water, and finally acetone again. The column is then dried by passing N_2 or air through it. If the column is properly prepared, the interior wall should have a mirrorlike finish. The column is next fitted with a zero dead volume connector containing a frit, the porosity of which must be at least one-half the particle size of the packings.

Packing materials with particle sizes larger than 20 μm are packed dry using traditional methods employed in GC. Microparticulate materials (5 and 10 μm) are invariably packed using the wet-packing procedure (slurry packing), and the technique used most frequently is the high-pressure balanced-density technique. The balanced-density technique is used for packing particles with a wide particle size distribution of the larger particles (10 μm). A typical slurry contains tetrabromoethane or methyl iodide admixed with a nonpolar modifier (e.g., perchloroethylene). The apparatus is shown in Fig. 35. In addition to the down-flow method, the "upward packing" procedure is also employed (21). If the sample cylinder is filled upward with the slurry, agglomerates will tend to fall to the bottom of the reservoir and will not be swept into the column. Most often, a constant-pressure air-driven pump capable of output pressures of 5000 to 1000 psi is employed, although constant flow pumps

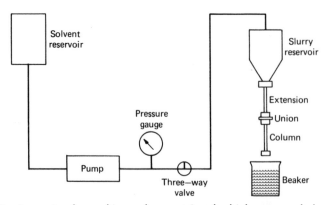

Figure 35 Apparatus for packing columns using the high-pressure balanced technique. Reproduced from reference 52 with permission.

may be used if their flow output is sufficiently high. The column is connected to a short, packed precolumn in order to maintain maximal packing density throughout the analytical column. In the absence of a precolumn, the density of the packed bed falls off sharply at the top of the column. The upward slurry-packing technique has also been used to minimize the timing requirements (21). Different slurry-packing procedures are summarized in Table 14.

A satisfactory slurry-packing procedure should fulfill the following requirements:

1   The rate of particle sedimentation during packing should be slow.
2   Agglomeration of particles should be avoided.
3   Pulsations during packing should be avoided.
4   Particles should be packed under high flow.
5   The slurry solvent should not react with the packing and it should be easily washed out of the packing.

Because of the large surface area of microparticles and high surface energy associated with electrostatic charges, microparticles agglomerate in the dry state and stick to the walls of the column. If a slurry is made

Table 14   Some Slurry-Packing Procedures[a]

Packing technique	Typical slurry solvent(s)	Reference
Balanced density	Tetrabromoethane, tetra-chloroethylene, diiodo-methane (with modifiers)	10,40,64 65,66
"Nonbalanced density"	Carbon tetrachloride	67,68
	Methanol	69,70
	Acetone	71
	Dioxane-methanol	72
	Tetrahydrofuran-water	73
	Isopropyl alcohol	74
	Chloroform-methanol	69
	Methanol-water-sodium acetete	75
Ammonia-stabilized slurry	0.001 M aqueous ammonia	76
High viscosity	Cyclohexanol, polyethylene glycol 200, ethylene glycol	77

[a]Reproduced (with modifications) from reference 10 with permission of Preston Publications Inc.

in an appropriate solvent, agglomeration will be eliminated. Furthermore, the solvent should have the same density as the packing (balanced-density solvent) in order to avoid size segregation during packing.

To prepare a column, a 10 to 15% (wt/v) slurry of the packing in a mixture of high-density halogenated hydrocarbons is prepared (14). This mixture is degassed and homogenized in an ultrasonic bath. If the solvent contains haloacids or elemental halogens (due to photodecomposition), they should be removed by passing the solvent through a silica gel column; otherwise, a drifting baseline or even cleavage of the organic modifier may occur. Several solvents have been proposed and used for the balanced-density slurry packing (1,5–9), and some are presented in Table 15.

Recent improvements that have resulted in narrower particle size distributions have lessened the need for matching the particle and slurry solvent densities. The difference in densities can be overcome by the

Table 15   Properties of Some Slurry-Packing Solvents*

	Density, ρ (g/ml)	Viscosity, η (cP, 20°C)
Diiodomethane (methylene iodide)	3.3	2.9
1,1,2,2-Tetrabromoethane[a]	3.0	–
Dibromomethane (methylene bromide)	2.5	1.0
Iodomethane (methyl iodide)	2.3	0.5
Tetrachloroethylene (perchloroethylene)[a]	1.6	0.9
Carbon tetrachloride[a]	1.6	1.0
Chloroform	1.5	0.6
Trichloroethylene	1.5	0.6
Bromoethane (ethyl bromide)	1.5	0.4
Dichloromethane (methylene chloride)	1.3	0.4
Ethylene glycol	1.11	1.7
Water	1.0	1.0
Pyridine	1.0	0.9
Tetrahydrofuran	0.9	0.5
η-Butanol	0.8	3.0
η-Propanol	0.8	2.3
Ethanol	0.8	1.2
Methanol	0.8	0.6
Cyclohexane	0.8	1.0
η-Heptane	0.7	0.4
Isooctane	0.7	0.5

[a]Most halogenated solvents are somewhat toxic, but these are particularly toxic.

*Reproduced (with modifications) from reference 36 with permission.

speed of the packing. Solvents such as acetone (20), isopropyl alcohol (13), methanol (2,10), and mixtures of methanol with dioxane (16) or chloroform (10) have been used for non-balanced-density slurry packing. Kirkland (6) reported the use of dilute aqueous ammonia solutions for spherical silica particles. However, the pH of the ammonia solution (10.1) exceeds the stability range of siloxane phases and thus limits the usefulness of this procedure for packing bonded phases.

With non-balanced-density slurry packing, it is extremely important that the solvent flow rate be as high as possible during the packing period in order to ensure bed compression. It has been suggested that alternative lowering and raising of the pressure (hammering) improves the homogeneity of the packing (5). When the column is packed, the pressure in the slurry reservoir should be released slowly. After the solvent flow has ceased naturally, the column should be released slowly; otherwise the packing may be disturbed.

An alternative method uses balanced-viscosity slurry solvents such as cyclohexanol, polyethylene glycol 200 and ethylene glycol (17,18), or a 20% solution of glycerol in methanol (19). These solvents prevent sedimentation, and in addition, they are also free from acidic impurities. However, due to their high viscosity, they offer high resistance to flow and necessitate extremely high packing pressures.

Columns of certain packing materials have to be eluted with appropriate solvents in order to remove the impurities present in the packing. Finally, the column should be conditioned with the solvent(s) to be used in the analysis.

6   ASSESSMENT OF COLUMN PERFORMANCE PARAMETERS

Before an LC column is put into operation, its performance must be tested. Prepacked columns should be tested for possible shipment damages and to establish whether the column performance complies with manufacturer's specifications. This can be done by chromatographing the manufacturer's test mixture under carefully reproduced conditions. The performance of homemade columns can be assessed using an appropriate test mixture. The standardized test procedures for determining both the kinetic and thermodynamic column performance have been discussed in the literature (3,22,23). The two basic tests of column performance include the determination of the plate count and peak symmetry. There are several methods available for measuring the column plate count. The most widely used method employs the following plate height equation:

$$h = \frac{H}{d_p} = \frac{1}{5.54}\left(\frac{L}{d_p}\right)\left(\frac{W_{\frac{1}{2}}}{t_R}\right)^2 = \frac{L}{Nd_p} \qquad (91)$$

where $W_{\frac{1}{2}}$ is the width of the peak at half-height. All other symbols have their usual meaning. The performance index, π, defined by the following equation (24), can also be used:

$$\pi = \frac{N^2}{(t_R\Delta p)} = \left(\frac{N}{t_R}\right)\left(\frac{N}{\Delta p}\right) \qquad (92)$$

where Δp is the pressure drop in bars or atmospheres.

In addition, the quantity known as the separation impedance (E) (25) represents an absolute way of determining column performance:

$$E = \frac{t_R\Delta p}{N^2\eta(1 + k')} = \frac{H^2}{K} \qquad (93)$$

where η is the mobile phase viscosity, and K is the chromatographic column permeability defined by:

$$K = \frac{\eta L^2}{\Delta p t_0} = \frac{\eta L u}{\Delta p} \qquad (94)$$

Typical performance values to be expected with microparticulate packings are approximately 65,000 plates/m.

It should be noted, however, that serious errors can result in the measurement of column plate count if the peaks are not Gaussian. Therefore, it is important to determine the peak asymmetry factor (AF), calculated at 10% peak height, as illustrated in Fig. 36, and defined by the following equation:

$$AF = \frac{\overline{CB}}{\overline{AC}} \qquad (95)$$

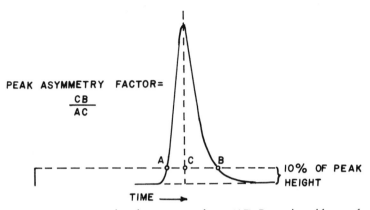

Figure 36   Determination of peak asymmetry factor (AF). Reproduced from reference 26 with permission of Preston Publications Inc.

Well-packed columns give asymmetry factors between 0.9 and 1.4. The peak asymmetry factor should not exceed 1.6. Since distortions of chromatographic peaks (non-Gaussian shape) can affect the N values, it is important to determine the extent of distortion. This can be achieved by means of the peak skew (27):

$$\text{Peak skew} = \gamma' = \frac{2(\tau/\sigma)^3}{[1 + (\tau/\sigma)^2]^{3/2}} \tag{96}$$

where σ is the standard deviation, and τ is the time constant for the exponential peak-tailing function. This parameter, derived from an exponentially modified Gaussian model, requires a computer simulation study.

In addition, the column flow resistance parameter (Φ) should also be calculated according to the following equation (22,28,29):

$$\phi = \frac{\Delta p d_p^2 t_0}{L^2} \tag{97}$$

For a well-packed column, ϕ should be in the range of 500 to 1,000.

7 COLUMN "INFINITE DIAMETER" AND WALL EFFECTS

Before Knox and Parcher (30) described the "infinite diameter effect," it was common practice to use 2-mm bore columns regardless of the particle size. When a sample is injected centrally into the chromatographic bed, if the column diameter is wide enough, radial dispersion of the solute will be relatively slow compared to the axial dispersion. Thus the solute band will traverse the entire column length without reaching the walls. However, if the solute is not injected at a point in the center of the column inlet, it will reach the column wall, and significant band dispersion will result due to the packing inhomogeneity and the concomitant flow variations in this region. Knox, Laird, and Raven (31) have shown that the dispersion in this region (which extends to approximately $30d_p$) is more rapid than in the center of the column. The extent of the wall region is inversely proportional as the ratio of column to particle diameter (d_c/d_p). The transverse distribution across the column is governed by Eq. 98:

$$W_r = 32D_m t_0 + 2.4Ld_p + W_i^2 \qquad (98)$$

where W_r is the radial peak width, t_0 is the time for an unretained peak, and W_i is the initial peak width at injection. The first term in Eq. 98 accounts for molecular diffusion and the second is for the effects of stream splitting. In order to achieve the infinite diameter effect and to predict how it varies with the reduced velocity, the column bore should be chosen according to the following requirement (31):

$$\frac{(d_c - 60d_p)^2}{Ld_p} > 16\frac{1.4}{(v + 0.060)} \qquad (99)$$

This means that if a point-injection technique is used, a column with a 5- to 7-mm bore packed with 5- or 10-μm particles will behave as if it had a very large diameter, and the solute will traverse the column length before it reaches the column wall. However, most columns do not operate in the infinite diameter mode. Usually, the sample and the mobile phase are injected through a small diameter connecting tubing onto a frit at the center of the column inlet (Fig. 37). With poorly packed columns or columns with an inadequate combination of particle size and column dimensions, the standard injection technique may result in significant band broadening due to the wall effects. This can be circumvented by using the point-injection technique with a syringe (Fig. 38) (26) or a microsampling valve (Fig. 39) (26), which enables central introduction of the sample while the mobile phase is directed across the entire inlet

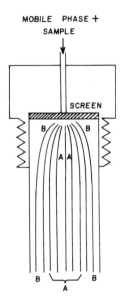

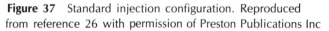

Figure 37 Standard injection configuration. Reproduced from reference 26 with permission of Preston Publications Inc

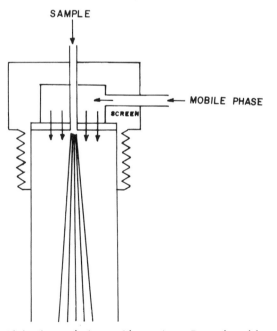

Figure 38 Point injection technique with a syringe. Reproduced from reference 26 with permission of Preston Publications Inc.

117

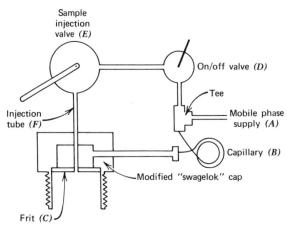

Figure 39   Point-injection apparatus with high-pressure microsampling valve. Reproduced from reference 26 with permission of Preston Publications Inc.

cross section. However, this requires special hardware and often results in reduced column capacity, since only a fraction of the column packing is used in the separation. With well-packed columns, standard injection techniques are permissible since the wall effects are usually not significant.

8 COMMERCIALLY AVAILABLE MICROPARTICULATE RP COLUMNS

RP bonded packings are available in three types: large porous, porous-layer beads (pellicular packings), and microparticulate packings. Large porous and porous-layer beads, which were the standard packing materials in the late 1960s and early 1970s, have been replaced by microparticulate packings and are now used mainly as guard column (precolumn) materials. Table 16 outlines the properties of HPLC packings.

    Presently, a large variety of microparticulate RP packings are commercially available. Among them, the ones with octadecylsilane functionality are most widely used. Table 17 summarizes the majority of packings currently in use. Phenylsilanes and nitrile phases are the most popular packings of intermediate polarity. They exhibit higher selectivity with polar compounds than octadecylsilanes, and they may be used in

Table 16   Properties of HPCL Packings*

Property	PLBs	Microparticles
Av. particle size, μm	30–40	5–10
Best HEPT[a] values, mm	0.2–0.4	0.01–0.03
Typical column lengths, cm	50–100	10–30
Typical column diameters, mm	2	2–5
Pressure drop, psi/cm[b]	3	20
Sample capacity, mg/g	0.05–0.1	1.5
Surface area (LSC), sq m/g	10–15	400–600
Bonded phase coverage, % wt	0.5–1.5	5–20
Ease of packing	easy, dry pack	difficult, slurry pack

[a]HEPT = height equivalent to a theoretical plate.
[b]Columns of equal dimensions (2.1 mm id) operated at linear velocity of
1 cm/sec and mobile phase viscosity of 0.3 cp.
*Reproduced (with modifications) from reference 15 with permission.)

"normal" as well as the RP modes. In addition, some packings with polar groups, for example, the amino functionality in its protonated form, can also function as anion exchangers.

The mean particle diameters, d_p, range between 5 and 10 μm. Most often, the nominal d_p value is specified, although some suppliers also designate the standard deviation (± 10 to 20%) or the size distribution (d_{p10}, d_{p50}, and d_{p90}). Several methods have been used for the determination of particle sizes: microscopy, sedimentation, electrical sensing zone method, wet sieving, and photosedimentation (37). Since each method is based on different assumptions, the particle size values obtained for the same packing will differ (37).

Although several theoretical studies (28,38) have shown that particle sizes of less than 5 μm would give higher efficiencies, relatively little is published on the performance of packings with 3- to 4-μm particles (39,40). This is due to the experimental problems associated with packing small particles, such as decrease in permeability, increase in temperature due to friction, the extracolumn effects, and the relatively large time constant of the detector and recorder (41). The phase loading varies from 5 to 22%; packings with low loading are indicative of homogeneous, monomeric phases.

In spite of the increasing popularity of homemade columns, commercial prepacked columns still find widespread use. This is mostly due to the belief held by many researchers that the credibility and utility of developed assays will be greater if separations are performed on prepacked, com-

Table 17 Some Microparticulate Packings and Prepacked Columns for Bonded Phase Chromatography*

Type (based on functional group)	Name	Supplier(s)	Form[a]	Functionality	Base material	size/ μm	Description
Non-polar RP	LiChrosorb RP-18	E. Merck	B or C	octadecylsilane	LiChrosorb	5, 10	monolayer, 22% loading
	μBondapak-C$_{18}$	Waters	C	octadecylsilane	Porasil	8-12	10% loading
	MicroPak-CH	Varian	B or C	octadecylsilane	LiChrosorb Si60	10	polymeric layer, 22% loading
	Nucleosil-C-18	Macherey-Nagel	B or C	octadecylsilane	Nucleosil	5, 10	capacity twice tha of C-8
	VYDAC RP-TP	Separations Group	C	octadecylsilane	VYDAC TP	10	10% loading
	ODS SiL-X-1	Perkin-Elmer	B or C	octadecylsilane	SiL-X-1	13 ± 5	irregular shape
	Partisil-ODS-2	Whatman	C	octadecylsilane	Partisil	10	16% loading
	Spherisorb-ODS	Phase Sep	B or C	octadecylsilane	Spherisorb	5, 10	spherical
	Zorbax-ODS	DuPont	B or C	octadecylsilane	Zorbax SiL	5-7	spherical, monofunctional, 22% loading

Polarity	Column	Manufacturer	Type[a]	Functional group	Base material	Particle size	Comments
Medium polarity	Allylphenyl Sil-X-1	Perkin-Elmer	B or C	-allylphenyl	SiL-X-1	13 ± 5	
	μBondapak-CN	Waters	C	alkylnitrile	Porasil	8-12	9% loading
	MicroPak-CN	Varian	C	alkynitrile	LiChrosorb S160	10	prep, Columns available
	Nucleosil C-8	Macherey-Nagel	B or C	octysilane	nucleosil	5, 10	general purpose RP
	Vydac TP Polar	Separations Group	B or C	alkylnitrile	Vydac TP silica	10	
High Polarity	BondaPak-NH$_2$	Waters	C	alkylamine	Porasil	8 – 12	9% loading
	MicroPak-NH$_2$	Varian	C	aminopropyl	LiChrosorb S160	10	prep, columns available
	Nucleosil-N(CH$_3$)$_2$	Macherey-Nagel	B or C	dimethylamine	Nucleosil	5, 10	Irregularly shaped 6 μ/m²

a B = bulk packing only; C = packed columns only.

*Reproduced (with modifications) from reference 52 with permission.

121

mercially available columns of guaranteed performance. Usually, pre-packed columns are supplied with a printout of the test run, including parameters such as k', HETP, peak asymmetry factor, and so on. Since the test mixtures are simple, it is easy to retest the column in order to check the specifications. It is useful to repeat this procedure periodically during use of the column, in order to monitor its state and determine if any deterioration has taken place. It should be pointed out that malfunctioning of the injector port, leakage in the connectors, and so on will have deleterious effects on the column performance and may often disguise the cause of the problem.

Most analytical prepacked columns are available in sizes of 25- to 30-cm length by 4- to 4.6-mm I.D. Although it has been shown that wider bore columns generally afford higher efficiency (infinite diameter effect) and higher loading capacity, short columns of narrower internal diameters are also being used. Their main advantage is decreased peak dilution, which is advantageous in trace analysis. The main barrier to improvements in detection limits resulting from the use of short columns is the extra-column broadening (injector and detector) which will no longer be over-shadowed by the column effects.

Recent studies have indicated that microbore columns (1-m long by 1-mm I.D.) may be an attractive alternative to the standard columns in terms of efficiency, speed of analysis, and low solvent consumption (32–34). Originally introduced by Horváth and Lipsky (35), these columns have been replaced by wider bore and shorter columns, primarily due to the effects of band broadening resulting from the design of various LC components (detector cells, injection device, pumps) which made their use impractical. However, it has been demonstrated (32) that by modifying the standard HPLC equipment (reducing the detector cell volume to 1 μl and adapting the pumps to deliver a flow of 2 to 100 μl/min), high efficiencies (up to 750,000 plates) and very rapid analyses can be achieved with microbore columns. It is doubtful, however, that these columns will find immediate use in routine work due to the need for equipment modification.

9 GUARD COLUMNS

In order to avoid contamination and prolong the life of analytical columns, particularly in biomedical or biochemical work, guard columns are installed between the injection system and the analytical column. These guard columns are short (usually 5 to 10 cm) and can be dry-packed with pellicular packing analogous in composition to the analytical column. If the system is properly engineered, the plate count of the analytical column will not be significantly decreased. Marked decrease in the plate count is observed only for peaks with $k' = 0$ (unretained on both the guard column and analytical column) (36). However, this does not have any practical significance, since the peaks of interest should not be unretained if the operating parameters are properly chosen.

10   OPERATION OF MICROPARTICULATE RP COLUMNS

In order to achieve optimal column performance and prolong the life of microparticulate packings, considerable care is required during their use and storage. This is especially important since commercially available prepacked columns are rather costly. The stability of an RP column depends on the sample preparation technique and the purity of the solvents used. RP columns are often used for the analysis of biological samples whose components are not always completely eluted, which gives rise to chemical contamination of the column. In addition, physiological samples must be precleaned to remove proteins and particulate matter that can impair the performance by irreversible adsorption and/or clogging of the column. Excessive flow rates can cause crushing of silica particles, leading to a pressure buildup in the column.

Generally, poor performance of microparticulate columns can be caused by three factors:

1   Physical degradation of the packing.
2   Chemical degradation.
3   Chemical contamination.

Improperly packed RP columns or well-packed columns used with very high flow rates can develop channeling (voids). Channeling can seriously degrade column performance by causing excessive on-column band dispersion. If the void develops at the head of the column, additional packing can be added using either the dry-packing or the slurry method. Dry-packing or tap-fill is usually more successful with particles of 20-μm diameter or larger. Microparticulate columns (5–10 μm) should be refilled using the slurry method, which involves filling of the void with methanol and adding small increments of the packing. After each successive addition, the packing should be allowed to settle. The process is repeated until the void is completely filled. The top is then smoothed with a flat spatula. The head of the column should be visually inspected after the solvent has been pumped through the column. If channeling develops throughout the chromatographic bed, double peaks develop. It is not possible to correct channeling without repacking the column.

Alteration of the basic chemical nature of either the support or the bonded phase, which occurs only under extreme conditions, results in permanent degradation of the column, and it is virtually impossible to return the column to its original condition. With silica-based columns, alteration of the support can occur at pH values below 2 and above 8. In such cases, the packing has to be discarded and the column repacked.

However, permanent chemical modification occurs only rarely if care is taken to avoid highly acidic and highly basic solvents. Packings contaminated by strongly adsorbed contaminants from impure solvents or incompletely eluted samples can be successfully reactivated. Chemical contamination will be manifested by decreased k' values and loss in resolution because of adsorption of the impurities onto the active sites of the packing. Column manufacturers suggest different procedures for reactivation of contaminated columns. Octadecyl or octyl, phenyl, cyano, and fatty acid columns can be washed with 5 to 10 column volumes of pure methanol during which several 2-μl aliquots of dimethyl sulfoxide (DMSO) should be injected. Finally, the column is washed again with pure methanol. During the DMSO wash, a sharp rise in column back pressure is observed, which is caused by the high viscosity of DMSO, lipids, or other nonpolar substances. These contaminants can be removed by the following procedure: The column is initially washed with 5 to 10 column volumes of methanol, followed by methylene chloride and n-heptane wash. Finally, the column should be washed with approximately 50 ml of methylene chloride and methanol. Solvents used in column reactivation must be virtually miscible and should not have high viscosity; otherwise disturbance of the column bed will occur. In addition, it is important to avoid generating heat of mixing upon change of solvents, since this would cause gas bubbles in the mobile phase and formation of voids. It is also possible to reactivate RP columns thermally (maximum temperature, 150°) in a GC or LC oven. After the column has been eluted with a series of solvents of decreasing polarity, the temperature is raised slowly and the column is allowed to bleed overnight (37).

11 MOBILE PHASES FOR RPLC

Polar solvents and/or their mixtures are commonly used in RP separations, and their eluting strength is reversed (44) compared to the Snyder eluotropic series for alumina (45).

With the increased sensitivity of most HPLC detection devices, solvent purity is of paramount importance. This has led to the use of specially distilled, high-purity (spectro-grade) solvents with specified spectral and chemical characteristics. Most RP separations employ buffered water as the initial eluent. The use of buffers is particularly important in separating polar or ionizable compounds on RP columns, since a properly chosen pH can significantly alter the retention of these compounds through secondary equilibria. The concentration of the salt in the buffer solution should be relatively high in order to avoid asymmetrical peaks and band splitting due to the slow rate of protonic and other secondary equilibria. Table 18 lists the useful pH ranges of some buffer systems.

Properly chosen buffers should have the following characteristics:

1 Buffering capacity in the pH range of 2 to 8.
2 Optical transparence.
3 Compatibility with organic eluents.
4 Capacity for enhancing equilibration rates.
5 Potential for masking silanol groups on the surface of the adsorbent.

Acetate buffers are infrequently used since low column efficiency results from the formation of nonpolar complexes between the acetate ions and positively charged solutes (45). In addition, halides are avoided because of their detrimental effect on the stainless steel in liquid chromatographs (45). The most troublesome aspect of water is its impure nature (46). This is particularly obvious in trace analysis performed by gradient elution, where impurities from water concentrated at the head of the column are eluted when the solvent strength is increased. This may give rise to "ghost peaks," or, more often, it may cause baseline drift, which can seriously affect the detectability of the late eluting peaks. Usually the purity of distilled-deionized water stored in well-capped glass bottles will suffice. In extreme cases, special purification procedures may be necessary in order to remove specific contaminants: reverse osmosis, electrolysis, ozonolysis, chromatographic removal of impurities using a large-

Table 18   pH Control Chemicals[a]

pH Control chemicals

Buffer systems	pH Range
$H_3PO_4/KH_2PO_4/K_2HPO_4/KOH$	2–12
Acetic acid/Na acetate	3–6
Acetic acid/NH_4 acetate/NH_4OH	5–9
NH_4 bicarbonate/NH_4 carbonate/NH_4OH	8–10
Na bicarbonate/Na carbonate/NaOH	9–11
$H_3BO_3/Na_3BO_3/NaOH$	7–11

Acids, bases, salts

Perchloric acid

Phosphoric acid

Acetic acid

Ammonium acetate

Ammonium carbonate

Sodium bicarbonate

Sodium carbonate

Sodium hydroxide

[a]Reproduced from reference 47 with permission.

diameter column with activated alumina on silica, distillation from alkaline permanganate, and so on.

Purity requirements vary depending on the type of application. They are particularly stringent in fluorometry, where traces of organic compounds can cause a drift in the baseline. Generally speaking, the requirements are less demanding in isocratic elution. High baseline offsets are usually needed with impure solvents, but after a steady-state equilibration, the detector baseline will level off. In addition to the electronic compensation in isocratic operation, the solvent effect can be reduced or eliminated by using a solvent-filled reference cell.

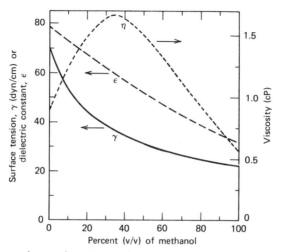

Figure 40 Dependence of pertinent solvent properties on the composition of water-methanol mixtures at 25°C. Surface tension, γ, data were obtained from Timmermans (12), and the data on viscosity, η, and dielectric constant, ε, are from Carr and Riddick (53) and Akerlof (54), respectively. Reproduced from reference 45 with permission of Preston Publications Inc.

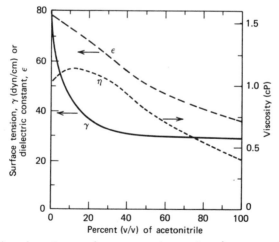

Figure 41 Plot of pertinent solvent properties against the composition of water-acetonitrile mixtures at 25°C. Data on surface tension, γ, and viscosity, η, are from Timmermans (12), and dielectric constants were obtained from Douheret and Morenas (12a). Reproduced from reference 45 with permission of Preston Publications Inc.

Among the organic solvents, acetonitrile and methanol, which are most commonly used in RP chromatography, have the highest eluting power or strength.

Physical properties of solvents such as viscosity, surface tension, and dielectric constant, which are important in determining the overall solvent selectivity, change as a function of composition (45–48). This is illustrated with water-methanol and water-acetonitrile mixtures in Fig. 40 and Fig. 41, respectively. Table 19 lists some properties of solvents most commonly used in RPLC separations.

Prior to the chromatographic analysis, solvents must be filtered through membrane filters in order to remove all particulate matter. Water is a particularly common source of this type of contamination, and the use of 0.5-μm filters in the mobile phase line is mandatory. In addition, samples injected into columns with particles <10 μm should be free of particulate matter. This is accomplished by filtration through Millipore membrane filters (0.5 μm) contained in a "Swinney" adapter.

In addition to the particulate impurities, dissolved gases are also present in the mobile phases, and each gas affects the chromatographic performance in a different way (42). The most commonly encountered problem is the occurrence of gas bubbles in the detector resulting from decreased gas solubility in the low-pressure components of the LC system. The result is detector noise (spikes) and baseline drift. In addition, the presence of gas bubbles in the pumps may cause variations in the flow due to differences in compressibility. The solubility of several gases in solvents of different polarities is shown in Fig. 42. The graph clearly demonstrates increased gas solubility in low-polarity solvents. The nonlinear gas solubility in mixed solvents such as those generated during a gradient run has also been described in the literature (51). It should be pointed out that contrary to the widespread belief, the solubility of bases increases with decreasing solvent polarity.

Dissolved oxygen is particularly undesirable since it forms UV-absorbing complexes with many solvents (51). This effect is pronounced at wavelengths below 260 nm; therefore, changes in oxygen concentration during a gradient run cause an upward baseline drift in UV absorption. Chemical contaminants, which may be present even in unopened bottles, may sometimes require elaborate purification steps. Table 20 lists the sources of some contaminants and suggests the method for their removal. Dissolved oxygen is known to have undesirable effects on fluorescence measurements, due to quenching (50). The extent of quenching varies

Table 19 Some Useful Solvent Properties of Commonly Used RPLC Solvents*

	MW	B.P. (°C)	n^a	Uv^b (nm)	ρ^c (g cm^{-1})	η^d (cP)	ε^e	μ^f (Debye)	γ^g dyn cm^{-1}	E^h
Acetone[i]	58.1	56	1.357	330	0.791	0.322	20.7	2.72	23	0.56
Acetonitrile	41.0	82	1.342	190	0.787	0.358	38.8	3.37	29	0.65
Dioxane	88.1	101	1.420	215	1.034	1.26	2.21	0.45	33	0.56
Ethanol	46.1	78	1.359	205	0.789	1.19	24.5	1.68	22	0.88
Methanol	32.0	65	1.326	205	0.792	0.584	32.7	1.66	22	0.95
iso-Propanol	60.1	82	1.375	205	0.785	2.39	19.9	1.68	21	0.32
n-Propanol	60.1	97	1.383	205	0.804	2.20	20.3	1.65	23	0.82
Tetrahydrofuran	72.1	66	1.404	210	0.889	0.51	7.58	1.70	27.6	0.45
Water	18.0	100	1.333	170	0.998	1.00	78.5	1.84	73	0.45

[a] Refractive index at 25°C
[b] Uv cut-off; the wavelength at which the optical density of a 1 cm thick neat sample is unity as measured against air
[c] Density at 20°C
[d] Viscosity at 20°C
[e] Dielectric constant
[f] Dipole moment
[g] Surface tension
[h] Elutropic value on alumina according to Snyder (45)
[i] Not suitable for use with UV detector
*Reproduced from reference 45 with permission of Preston Publications, Inc.

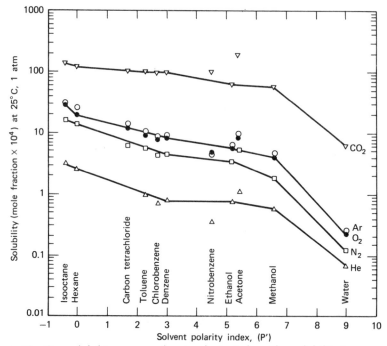

Figure 42 Gas solubility versus solvent polarity index. The solubility is expressed as mole fraction $\times$ 10^4 for pure gases in equilibrium with the solvent at 1 atm and 25°C. The polarity index is the Snyder parameter, ∇, CO_2; $\circ$, Ar; $\bullet$, O_2; $\square$, N_2; $\triangle$, He. Reproduced from reference 51 with permission.

with the type of solute analyzed, and it is particularly pronounced with aromatic hydrocarbons, aliphatic aldehydes, and ketones. In addition, baseline drifts are often encountered due to quenching of the background fluorescence of the mobile phase. Two other effects of dissolved gases include changes in refractive index and pH of unbuffered solvents with a change in the concentration of dissolved carbon dioxide.

In order to eliminate or reduce detector artifacts and oxidative degradation of samples, dissolved gases must be removed from the solvent system. This can be achieved by vacuum degassing, heating, and ultrasonic treatment. If heating is used, considerable losses of the more volatile component will occur unless the mixture is refluxed.

Recently, a new method of solvent degassing, known as purging or

Table 20   Solvent Purity in HPLC[a]

CONTAMINANT	POSSIBLE SOURCE	EFFECT	REMOVAL
Particulate matter (dust, etc.)	during transfer, unclean vessels	may block in-line filters, lodge in pump seals, or accumulate at column head	filter through membrane filter
Water	glassware, solvent preparation or manufacture	variable column activity, k' variation, stability of silicate ester bonded phases	dry over molecular sieve or anhydrous sodium sulfate.
Alcohol	stabilizer in $CHCl_3$, impurity in hydrocarbons	similar to water	from hydrocarbons, pass through activitated silica; from $CHCl_3$, extract with water, dry with Na_2SO_4
Hydrocarbons (in water)	organic matter	baseline instability during gradient elution	pass through porous polymer column or C_{18} bonded phase
Peroxides (in ethers)	degradation	oxidation of bonded phase (e.g., $- NH_2$ to $- NO_2$), reaction with sample, column deactivation or degradation (polystyrene-based)	distill or pass through activated silica gel or alumina
HCl, HBr (halogenated solvents)	degradation	column degradation, expecially bonded phases, UV absorbance (bromide), stainless steel attack	pass through activated silica or $CaCO_3$ chips
BHT, hydroquinone	antioxidants in THF	UV - absorbing	distill
Dissolved oxygen	solvent preparation	degrades polystyrene-based packings, oxidizes B,B'-oxydipropionitrile, may react with sample	degas solvent with vacuum or heat
Unknown UV-absorbing	from manufacture	baseline instability or drift during gradient elution, high detector background	use activated silica or alumina, or distill for organics, recrystalize or pass over ion exchange column for inorganics
High boiling compounds	from solvent manufacture	contaminates collected sample in preparative HPLC	distill
Algae in water	growth during prolonged storage	can plug in-line filters, column entrance frits	distill from alkaline permanganate or discard.

[a]Reproduced from reference 15 with permission.

stripping, has been introduced (48,51). This technique involves bubbling of helium (or neon) through the solution, which results in the elimination of all gases except helium from the mobile phase (44). Although at present there is no information on the solubility of helium in mixed solvents, two possible explanations have been advanced to explain the mechanism of this degassing technique (51). According to the first one, the solubility of helium may be nonlinear, but so small that the microbubbles formed do not pose any problems. An alternative explanation is that the solubility may be linear and appreciable so that only small amounts of helium may exceed the saturation level (51).

REFERENCES

1. R. E. Majors, *Anal. Chem.*, **44**, 1722 (1972); **45**, 755 (1973).
2. P. A. Bristow, *J. Chromatogr.*, **131**, 57 (1977); **149**, 13 (1978).
3. I. Molnár, *Chromatographia*, **12**(6), 371 (1979).
4. I. S. Krull, M. H. Wolf, and R. B. Ashworth, *Am. Lab.*, **10**, 45 (1978).
5. J. J. Kirkland, *Chromatographia*, **8**, 661 (1975).
6. J. J. Kirkland, *J. Chromatogr. Sci.*, **9**, 206 (1971); **10**, 593 (1972).
7. W. Strubert, *Chromatographia*, **6**, 50 (1974).
8. R. M. Cassidy, D. S. LeGay, and R. W. Frei, *Anal. Chem.*, **46**, 340 (1974).
9. J. J. Kirkland and P. E. Antle, *J. Chromatogr. Sci.*, **15**, 137 (1977).
10. R. E. Majors, *J. Chromatogr. Sci.*, **15**, 334 (1977).
11. C. J. Little, A. P. Dale, D. A. Ord, and T. R. Marten, *Anal. Chem.*, **49**, 1311 (1977).
12. J. Timmermans, *The Physico-Chemical Constants of Binary Systems in Concentrated Solutions*, Vol. 4, Wiley–Interscience, 1960.
12a. G. Douheret and M. Morenas, *Compt. Rend.*, **43C**, 729 (1967).
13. S. H. Chang, K. M. Gooding, and F. E. Regnier, *J. Chromatogr.*, **125**, 103 (1976).
14. J. J. DeStefano and H. C. Beachell, *J. Chromatogr. Sci.*, **8**, 435 (1970).
15. R. E. Majors, *J. AOAC*, **60**, 186 (1977).
16. H. R. Linder, H. P. Keller, and R. W. Frei, *J. Chromatogr. Sci.*, **14**, 234 (1976).
17. J. Asshauer and I. Halász, *J. Chromatogr. Sci.*, **12**, 139 (1974).
18. R. Endele, I. Halász, and K. Unger, *J. Chromatogr.*, **99**, 377 (1974).
19. C. F. Simpson, Ed., *Practical High Performance Liquid Chromatography*, Heyden, London, 1976, p. 295.
20. G. B. Cox, C. R. Liscombe, M. J. Slucutt, K. Sudgen, and J. A. Upfield, *J. Chromatogr.*, **117**, 269 (1976).
21. P. A. Bristow, *LC in Practice*, HETP Publ., 10 Langley Drive, Handforth, Wilmslow, Cheshire, U.K., 1976, p. 333.
22. P. A. Bristow and J. H. Knox, *Chromatographia*, **10**(6), 279 (1977).
23. W. E. Hammers, R. H. A. M. Janssen, A. G. Baars, and C. L. DeLigny, *J. Chromatogr.*, **167**, 273 (1978).
24. M. Golay, in *Gas Chromatography 1958*, D. H. Desty, Ed., Butterworths, London, 1959, p. 36.
25. J. C. Giddings, *J. Chromatogr.*, **13**, 301 (1964).
26. J. J. Kirkland, W. W. Yau, H. J. Stoklosa, and C. H. Dilks, Jr., *J. Chromatogr. Sci.*, **15**, 303 (1977).
27. E. Grushka, *Anal. Chem.*, **44**, 1733 (1972).
28. J. H. Knox and M. Salem, *J. Chromatogr. Sci.*, **7**, 614 (1969).
29. J. H. Knox, *Ann. Rev. Phys. Chem.*, **24**, 29 (1973).

30. J. H. Knox and J. F. Parcher, *Anal. Chem.*, **41**, 1599 (1969).
31. J. H. Knox, G. R. Laird, and P. A. Raven, *J. Chromatogr.*, **122**, 129 (1976).
32. R. P. W. Scott and P. Kucera, *J. Chromatogr.*, **169**, 51 (1979).
33. D. Ishii, K. Asai, K. Hibi, T. Jonokuchi, and M. Nagaya, *J. Chromatogr.*, **144**, 157 (1957).
34. D. Ishii, K. Hibi, K. Asai, and T. Jonokuchi, *J. Chromatogr.*, **151**, 147 (1978).
35. C. Horváth and S. R. Lipsky, *Anal. Chem.*, **41**, 1227 (1969).
36. L. R. Snyder and J. J. Kirkland, *Introduction to Modern Liquid Chromatography*, Wiley-Interscience, New York, 1979, p. 229.
37. K. K. Unger and M. G. Gimpel, *J. Chromatogr.*, **180**, 93 (1979).
38. M. Martin, C. Ron, and G. Guiochon, *J. Chromatogr.*, **110**, 213 (1975).
39. I. Halász, H. Schmidt, and P. Vogtel, *J. Chromatogr.*, **126**, 19 (1976).
40. J. C. Kraak, H. Poppe, and F. Smedes, *J. Chromatogr.*, **122**, 147 (1976).
41. K. K. Unger, W. Messer, and K. F. Krebs, *J. Chromatogr.*, **149**, 1 (1978).
42. I. S. Krull, U. Goff, and P. B. Ashworth, *Am. Lab.*, **10**, 31 (1978).
43. K. Karch, I. Sebastian, I. Halász, and H. Engelhardt, *J. Chromatogr.*, **122**, 171 (1976).
44. L. R. Snyder, *Principles of Adsorption Chromatography*, Dekker, New York, 1968, pp. 194–195.
45. C. Horváth and W. Melander, *J. Chromatogr. Sci.*, **15**, 393 (1977).
46. R. Majors, *Varian Instrum. Appl.*, **10**, 8 (1976).
47. S. R. Bakalyar, *Am. Lab.*, **10**, 43 (1978).
48. S. R. Bakalyar, R. McIlwrick, and E. Roggendorf, *J. Chromatogr.*, **142**, 353 (1977).
49. F. W. Karasek, *Res. Dev.*, **28**, 38 (1977).
50. M. Furst, M. Kallman, and F. H. Brown, *J. Chem. Phys.*, **26**, 1321 (1953).
51. S. R. Bakalyar, M. P. T. Bradley, and R. Honganen, *J. Chromatogr.*, **158**, 277 (1978).
52. C. F. Simpson, Ed., *Practical High Performance Liquid Chromatography*, Heyden, London, 1976.
53. C. Carr and J. A. Riddick, *Ind. Eng. Chem.*, **43**, 692 (1951).
54. G. Akerlof, *J. Am. Chem. Soc.*, **54**, 4125 (1932).

BIBLIOGRAPHY

Bristow, P. A., *LC in Practice*, HETP Pub., 10 Langley Drive, Handforth, Wilmslow, Cheshire, U.K., 1976. p. 62.
Bristow, P. A., *LC in Practice*, HETP Publ., 10 Langley Drive, Handforth, Wilmslow, Cheshire, U.K., 1977.

Bristow, P. A., P. N. Brittain, C. M. Riley and B. F. Williamson, *J. Chromatogr.*, **131**, 57 (1977).

Bristow, P. A., and J. H. Knox, *Chromatographia*, **10**, 279 (1977).

Colin, H., and G. Guichon, *J. Chromatogr.*, **141**, 289 (19770.

Cooke, N. H. C., and K. Olsen, *Am. Lab.*, 45 (August 1979).

Coq, B., C. Gonnet, and J. L. Rocca, *J. Chromatogr.*, **106**, 249 (1975).

Cox, G. B., *J. Chromatogr. Sci.*, **15**, 385 (1977).

Grushka, E., *Bonded Stationary Phases in Chromatography*, Ann Arbor Science Publishers, Inc., Ann Arbor, MI, 1974.

Horváth, C. and W. Melander, *J. Chromatogr. Sci.*, **15**, 393 (1977).

Karger, B. L., and E. Sibley, *Anal. Chem.*, **45**, 740 (1973).

Knox, J. H., *J. Chromatogr. Sci.*, **15**, 352 (1977).

Kraak, J. C., H. Poppe, and F. Smedes, *J. Chromatogr.*, **122**, 147 (1976).

Linder, H. R., H. P. Keller, and R. W. Frei, *J. Chromatogr. Sci.*, **14**, 234 (1976).

Locke, D. C., *J. Chromatogr. Sci.*, **11**, 120 (1973).

Majors R. E., in *Practical High-Performance Liquid Chromatography*, C. F. Simpson, Ed., Heyden, London, 1976, Chap. 7.

Martin, M. and G. Guiochon, *Chromatographia*, **10**, 194 (1977).

Regnier, F. E., and R. Noel, *J. Chromatogr. Sci.*, **4**, 316 (1976).

Rehak, V., and E. Smoklova, *Chromatographia*, **9**, 219 (1976).

Unger, K. K., *Porous Silica: Its Properties and Use as a Support in Column Liquid Chromatography*, Elsevier, Amsterdam, 1979.

Webber, T. J. N., and E. H. McKerrell, *J. Chromatogr.*, **122**, 243 (1976).

V Separation Mechanisms

Since the development of chemically bonded RP packings, the separation mechanism has been a subject of continuing controversy. This is mostly due to the difference in the physiochemical basis for separation on polar and nonpolar phases. With polar phases, the predominant factor in determining retention is the interaction between the solute and the stationary phase; however, with nonpolar phases, the selectivity is mainly determined by the solvent effects. The interactions of nonpolar solutes with nonpolar stationary phases are rather weak and nonselective (dispersion forces of the van der Waals type). Although the nature of the organic group controls the selectivity with respect to the solute functional groups (1), the separation of closely related solutes is governed by the solution phenomena in the mobile phase (2).

Solute interaction in RP chromatography may proceed by three essentially different mechanisms:

1 Partitioning between the hydrocarbon layer on the surface of the packing and the mobile phase (LLC). This mechanism was observed with polymeric phases (3,7).

2 Partitioning between the mobile phase and the "modified" stationary phase. This occurs when the organic modifier is adsorbed on the highly active sites of the packing (4,8a).

3 Adsorption of solutes on the hydrocarbon surface layer (LSC) (1,2,5–8).

According to Pryde (9), "it seems a little irrelevant to argue whether the mechanism is by partition or by adsorption for neither term is strictly applicable." Several other authors (9–12) consider the mechanism to be a mixture of adsorption and partition. A comprehensive review of the subject is available in the literature (13).

Prior to the discussion of the possible mechanism of interaction, differentiation must be clearly made between adsorption and partition. Adsorption refers to the process in which solute and solvent molecules

135

compete for the active sites on the surface of the adsorber. The interfacial region (adsorbed surface-bulk mobile phase), where this interaction takes place, may be a monolayer or multilayer. The separation surface is arbitrarily chosen in an area in which the Gibbs surface excess for the solvent is zero.

In partition, the interfacial region is usually neglected, and the interaction occurs by partitioning of the solute between the mobile phase and the stationary phase coated on the support. Generally, no definite conclusions can be drawn from the standard methods for making a distinction between adsorption and partition: (a) the variation in k' with temperature or carbon number and the use of isotherms determined by a static method (14); (b) comparison of selectivity data obtained from a partitioning system and an analogous bonded RPLC system (12).

The main problem in distinguishing between the two possible mechanisms results from insufficient knowledge of the exact surface topology of the bonded phases. The bonded phase anchored to the surface of the support can be pictured as an alkyl "brush" (15) or "molecular fur" (16), with individual bristles making up the active surface. Alternatively, it can be envisioned as a hydrocarbonaceous sheath made up of recumbent alkyl chains which have aggregated to reduce the contact area with the polar solvent (17). The latter liquidlike structure is assumed to be composed of "liquid droplet clusters" which would permit a three-dimensional interaction with solutes by allowing the intercalation of solute molecules between the clustered alkyl chains (12). It is highly improbable, however, that the monomeric (non-cross-linked) bonded supports of the "brush type" will behave like a bulk liquid, since the average distance between the hydrocarbon chains is large and the bonded groups are deprived of some translational and relational degrees of freedom compared to the nonbonded state (13). However, Löchmuller et al. (12) stated that the isolated alkyl "bristles" would be energetically unfavorable and that hydrophobic association of bonded ligands is likely to occur in polar organic-aqueous solvents. This is in agreement with the observations of Frank and Evans (17), who suggested that the hydrophobic association of alkanes at low concentrations was the reason for a large negative entropy.

Polymeric phases are generally considered to act as an inert gel which contains small pools of mobile phase of different thermodynamic activity from the interstitial liquid (3,18). Therefore, some authors believe that because of the differences in solute solubility between these two regions, the predominant separation mechanism on these phases is partition, although adsorption may also occur. This is exemplified in increasing k' values with an increasing degree of surface coverage if the specific surface area is constant. In summary, these phases, with hydrocarbon bristles less than several thousand angstroms long, do not have the typical properties of either a partitioning liquid or a chemical adsorbent (18,19).

1 SOLVOPHOBIC THEORY

To describe the selectivity in RPLC, *solvophobic interactions* are often invoked. According to this model, completely or partially hydrophobic solute molecules are excluded from the polar solvent into the hydrocarbonaceous bonded phase. The solvophobic theory provides a framework for the treatment of phenomena associated with the properties of the mobile phase and assumes an ideal hydrocarbonaceous stationary surface (20,21). It has, nevertheless, shed light on the mechanism of RPLC interactions and thus contributed significantly to the general understanding of RPLC.

The interaction between the solute molecule and the nonpolar hydrocarbon surface depends on the weak, nonspecific dispersion forces (1, 5,16). The thermodynamic equilibrium constant for the binding process (K) can be expressed in terms of solute retention (k' value) and the volume ratios of the stationary and mobile phases (ϕ):

$$k' = \phi K \tag{100}$$

When a nonpolar solute or a solute with nonpolar segments is introduced into a polar environment, the dissolution process will be accompanied by a positive free energy (ΔG) as a result of a large negative entropy. To overcome this loss in entropy, structural ordering of the nonpolar solute segments in a polar solvent will take place, so that a cavity of polar solvent will be formed around the solute molecule (22). These effects, although not restricted to aqueous media, are very pronounced in water. The thermodynamics of the hydrophobic or, more generally speaking, solvophobic interactions have been developed by Sinanoğlu et al. (23,24) and adapted by Horváth et al. (5,25) to apply more specifically to the RPLC systems. According to this theory, hydrophobic interactions result from repulsive forces between a polar solvent and the nonpolar solute and stationary phase. The driving force in the binding of the solute to the stationary phase is the decrease in the area of the nonpolar segment of the solute molecule exposed to the solvent. Stated differently, it is the *repulsion* of the solute and the polar mobile phase, rather than the weak nonpolar interactions between the solute and the nonpolar stationary phase, that is responsible for the association of the solute and stationary phase.

Hydrophobic interactions have attracted considerable attention after the discovery that they play a significant role in the three-dimensional structure of many biologically active polymers (26,27), for example, in the structure and flexibility of membranes. The theoretical treatments involve either the study of thermodynamics of solute transfer between water and a nonpolar medium or the statistical thermodynamic treatment of aqueous solutions. However, these theories are not directly applicable to the RP systems in which the mobile phase is rarely pure water. The modified approach assumes that the nonionic solute *(S)* binds to the hydrocarbon ligand *(L)* in a reversible reaction, forming a complex *(SL)*:

$$S + L \rightleftharpoons SL \tag{101}$$

The equilibrium constant for this reaction can be expressed as:

$$K = \frac{[SL]}{[S]\,[L]} \tag{102}$$

K is related to the free energy change through the following relationship:

$$\ln K = \frac{-\Delta G°}{RT} \tag{103}$$

where R is the universal gas constant and T is the absolute temperature.

The retention of the solute can be expressed in terms of the overall energy change for the binding of the solute to the stationary phase:

$$\ln k' = \ln \phi - \left(\frac{\Delta G°}{RT}\right) \tag{104}$$

The free energy change accompanying formation of the complex *SL* can be thought of as being composed of two discrete hypothetical processes:

1 Formation of the complex SL in the gaseous phase.

2 Interaction of the complex SL with the solvent.

Thus the total free energy change is given by:

$$\Delta G^\circ_{\text{solv assoc}} = \Delta G^\circ_{\text{gas assoc}} + \Delta G^\circ_{\text{net solv effect}} \qquad (105)$$

The association in the gaseous phase is assumed to occur through the van der Waals interactions, and the corresponding free energy change is denoted by ΔG_{assoc}. The second process is more difficult to deal with, and the change in free energy associated with it ($\Delta G^\circ_{\text{net solv effect}}$) is composed of the sum of the free energy differences for the formation of solvent cavity (ΔG_{cav}), the interaction of the complex with the solvent, and the statistical term representing change in free volume:

$$\Delta G^\circ_{\text{net solv effect}} = \Delta G^\circ_{\text{net cav}} + \Delta G^\circ_{\text{net inter}} - RT \ln \left(\frac{RT}{P_0 V} \right) \qquad (106)$$

where $P_0 = 1$ atm, and V is the mole volume of the solvent. The second term in Eq. 106 (25) can be further broken down into two full energy contributions which arise from the van der Waals and electrostatic interactions, giving:

$$\Delta G^\circ_{\text{net solv effect}} = \Delta G^\circ_{\text{net cav}} + \Delta G^\circ_{\text{net vdW}}$$
$$+ \Delta G^\circ_{\text{net e.s.}} - RT \ln \left(\frac{RT}{P_0 V} \right) \qquad (107)$$

The individual terms in the overall solvent effect are graphically illustrated in Fig. 43.

The free energy of the van der Waals interaction between the solute and the solvent can be expressed as:

$$\Delta G^\circ_{\text{net vdW}} = -W - a\Delta A \qquad (108)$$

where W and a are parameters that are properties of the solvent, and ΔA is the reduction in molecular surface area upon formation of the solute–stationary phase complex (25).

In RP separations, the electrostatic free energy term $\Delta G^\circ_{\text{net e.s.}}$ depends

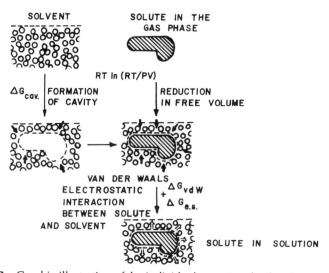

Figure 43 Graphic illustration of the individual terms involved in the solvent effect. Reproduced from reference 26 with permission.

on the charge distribution, molecular size of the solute, ligand, and complex, and the dielectric constant of the solvent, ε:

$$\Delta G^\circ_{\text{net e.s.}} = \frac{\Delta Z}{\varepsilon} \tag{109}$$

where $\Delta Z = Z_{SL} - Z_S - Z_L$.

The free energy change, calculated from the physical properties of the solute and solvent, can be used to predict the retention behavior of the compound if the solvophobic association is the only mode of interaction (25). The free energy change associated with the formation of a cavity within the solvent can be expressed by the following relationship:

$$\Delta G^\circ_{\text{net cav}} = - [N\Delta A + 4.8N^{1/3}(\kappa^e - 1)V^{2/3}]\gamma \tag{110}$$

where ΔA is the decrease in surface area upon formation of the complex, N is the Avogradio's number, κ^e is a factor for converting the bulk surface to the molecular dimensions, and γ is the surface tension of the bulk solvent.

From Eq. 110 it is evident that the free energy term for the cavity formation increases with increasing contact area between the solute and the stationary phase and with increasing surface tension of the solvent (25). Thus the extremely high surface tension of water results in an abnormally high $\Delta G°_{net\ cav}$ term in this solvent. Addition of an organic modifier such as methanol or acetonitrile and ionic salts affects the surface tension of water. The effect is illustrated in Fig. 44.

Taking into account all individual contributions, the overall standard free energy change accompanying solute retention can be expressed as follows:

$$\Delta G°_{assoc\ solv} = \Delta G°_{assoc\ gas} + \Delta G°_{net\ cav} + \Delta G°_{net\ vdW}$$
$$+ \Delta G°_{net\ e.s.} - RT \ln \left(\frac{RT}{P_0 V} \right) \quad (111)$$

Detailed theoretical treatment and the derivation of Eq. 111 expressed in terms of measureable physicochemical parameters can be found in the literature (25).

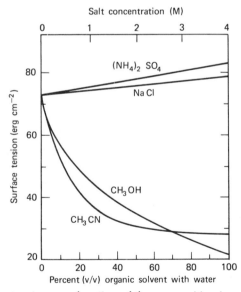

Figure 44 Surface tension as a function of the composition in mixed solvents and salt solutions. Water-methanol and water-acetonitrile systems represent the mixed solvents. Reproduced from reference 5 with permission.

By combining all free energy terms, an expression can be obtained that relates the logarithm of the capacity factor (k') to the free energy change associated with the formation of the complex:

$$\ln k' = \phi + \frac{1}{RT} [\Delta A(N\gamma + a) + NA_s\gamma(\kappa^e - 1) + W - \frac{\Delta Z}{\varepsilon}] \qquad (112)$$
$$+ \ln \frac{RT}{P_0V}$$

where A_s is the surface area of the solvent molecule, and all other terms have their usual meaning (25).

2 MOLECULAR CONNECTIVITY

To estimate the contributions of nonpolar solute segments to retention in RP separation, a new topological index, called the "molecular connectivity," χ, was introduced (28,31). This index is proportional to the cavity surface area and is defined by the following expression:

$$\chi = \sum_{k=1}^{k} \frac{1}{(\delta_i\delta_j)^{1/2}} \qquad (113)$$

where $\delta = 1, 2, 3$, or 4, corresponding to the number of atoms attached to atoms i and j, and k is the number of bonds in the solute molecule or its nonpolar segment. In calculating the χ values, only the skeleton of the compound is considered and the hydrogen atoms are neglected. It has been shown that in RPLC, log k' is proportional to log $S_{H_2O}(S_{H_2O} =$ solute solubility in water) and, in turn, to χ (22). Thus this relationship can be used to calculate the water solubility of nonelectrolytes and predict their retention. This method was used in estimating hydrophobic selectivity for certain classes of compounds (22).

3 EFFECT OF SOLVENTS IN RPLC

The solvophobic theory owes its current popularity to its ability to account for different phenomena operating in RP chromatography and predict the effect of change in separation parameters on the overall retention.

It has been stated before that the solute interactions with bonded re-versed phases are rather weak. If the same RPLC column is used in GC, the retention of a particular solute would be high, which explains the paramount importance of the mobile phase in RP separations (26). The retention in LC can be expressed as a sum of the solute–stationary phase interactions in GC and the adverse effect of the solvent on these inter-actions in LC. Lower retention in LC compared to GC can be explained in terms of decreased solute–stationary phase interactions in the liquid medium. The individual contributions of different terms in Eq. 107 to the overall solvent effect are depicted in Fig. 45. The experimental results were obtained for undissociated toluic acid on an ODS column with a water-acetonitrile solvent system (26). It is evident from Fig. 45 that the van der Waals energy for the solute-solvent interactions term is large. The contribution of the electrostatic interaction is small in this case; thus

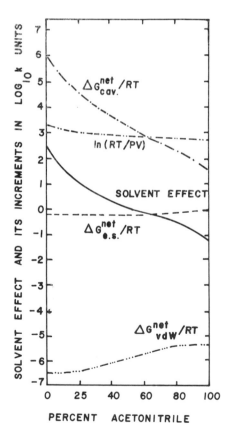

Figure 45 Solvent effect in RP chro-matography. The individual contribu-tions representing the terms of Eq. 107 are plotted as the function of the com-position of the water-acetonitrile eluent. The solute is undissociated toluic acid. Experimental results obtained with an octadecyl-silica column at room tem-perature were used to estimate the data presented here. The logarithm of the ca-pacity factor on the ordinate can be con-sidered as a dimensionless energy ap-propriate at the temperature of the experiment. Reproduced from reference 26 with permission.

its effect on the overall solvent effect is quite negligible. In the absence of the cavity term, the sum of the van der Waals and the free volume terms would result in marginal retention. Therefore, the "hydrophobic effect" is obviously dominated by $\Delta G°_{net\ cav}/RT$, the gain in energy associated with the decrease in cavity accompanying the formation of the complex. The two major eluent parameters that influence this term in the expression for the capacity factor are the surface tension and the dielectric constant (16). However, the effect of the dielectric constant on nonpolar solutes is small and it can be neglected (16). With ionized species, the effect of the dielectric constant of the medium is larger due to the secondary equilibria (ionization suppression).

The decrease in surface tension, resulting from an increase in concentration of the organic modifier in the eluent, leads to a decrease in retention, as predicted by Eq. 112. The influence of the two surface tension-

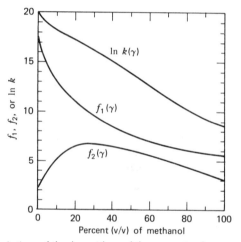

Figure 46 The variation of the logarithm of the capacity factor and that of the two surface-tension-dependent terms in Eq. 105 with the composition of water-methanol mixtures. The function $f(\gamma)$ is given by $N\Delta A\gamma/RT$, where ΔA is the surface area change that occurs upon binding of the solute by the nonpolar ligand of the stationary phase, and γ is the surface tension of the eluent. The value A is taken as 100 Å². The bottom curve, $f_2(\gamma)$, represents the term $NA_s(K - 1)/RT$, where A_s is the average surface area of the solvent molecules and κ^e converts the microscopic surface tension to molecular dimensions. The line at the top is $f_1(\gamma) + f_2(\gamma)$ and very nearly reproduces the solvent dependence of the logarithm of the capacity factor obtained from experiments with nonionized solutes. The surface tension was obtained from Timmermans (34), and the corresponding κ^e and A_s values are from Horváth et al. (5). Reproduced from reference 16 with permission of Preston Publications, Inc.

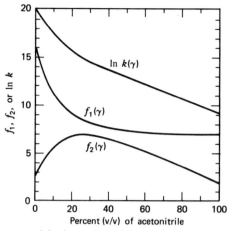

Figure 47 The variation of the logarithm of the capacity factor and the two surface-tension-dependent terms in Eq. 105 with the composition of water-acetonitrile mixtures. The definition of all terms and pertinent references are given in Fig. 45. Reproduced from reference 16 with permission of Preston Publications, Inc.

dependent terms on the logarithm of the capacity factor and their combined effects for water-methanol and water-acetonitrile mixtures are shown in Figs. 46 and 47, respectively. It should be pointed out that the magnitude of the slopes is dependent on the solute size and surface area (16).

For nonionized species, this quasi-linear relationship between the solvent composition and retention has been verified experimentally (5, 22,32). The linearity of the relationship thus enables further simplification of the prediction of capacity factors by means of an empirical formula:

$$k' = k'_0 \, e^{mx} \tag{114}$$

where k'_0 is the capacity factor in pure aqueous solvent, m is a constant for a given solute, and x is the percent (v/v) of the organic modifier in the mobile phase (16).

The solvophobic theory can successfully explain and predict the retention behavior of different classes of compounds. Thus, for example, the linear relationship between the logarithm of the capacity factor and the carbon number in a homologous series can be explained in terms of solute size and surface area: with increasing carbon number, the change in surface area upon binding and thus the free energy accompanying the formation of the complex are increased. This is reflected in increasing retention of successive members of the series (26). Upon change in solvent composition from aqueous buffer to methanol or acetonitrile, a linear decrease in $\log k'$ values with an increasing amount of the organic modifier is observed (26). The effect of solute size on retention is manifested in the slope of the plot of $\log k'$ versus solvent composition, which increases with increasing surface area. This results from the decrease in the solvent surface tension with an increasing percentage of the organic modifier and an increase of the surface-area change in larger molecules.

4   EFFECT OF ADDED SALTS

The solvophobic theory can also account for the effect of salts in the mobile phase on solute retention. The influence of salts can be explained in terms of reduced electrostatic repulsion between solute molecules and the increase in eluent surface tension, which can be deduced from the modified Debye-Hückel theory of dilute electrolytes (32) and Hofmeister (lyotropic) series (33), respectively.

The addition of neutral inorganic salts causes a linear increase in the surface tension of aqueous solutions, according to the following equation:

$$\gamma = \gamma_0 + \sigma m \tag{115}$$

where γ_0 is the surface tension of pure water, m is the molarity of the solution, and σ is the molal surface tension increment of the salt obtained from the Hofmeister series (26). The logarithm of the capacity factor of neutral solutes increases linearly with increasing salt concentration, following an increase in the eluent surface tension. The effect of added salt on retention of ionized molecules is more complex. The change in surface tension with the change in salt and/or organic modifier concentration is illustrated in Fig. 48.

The effect of changing surface tension on the logarithm of the capacity factor, provided κ^e and mole volume of the eluent remain constant, can be expressed by a simplified equation:

$$\ln k' = A''' + B'' \tag{116}$$

where A''' contains all the terms in the equation that are influenced by the surface tension, and B'' is the following:

$$B'' = \frac{N\Delta A + 4.836N^{1/3}(\kappa^e - 1)V^{2/3}}{RT} \qquad (117)$$

where all the terms have their usual meaning.

The change in the logarithm of the capacity factor with the change in salt concentration is illustrated in Fig. 48. It should be pointed out that with mixed solvents the equation does not hold, since the capacity factor decreases with decreasing surface tension in a nonlinear fashion. A detailed explanation of the influence of mixed solvents on solute retention is available in the literature (5).

5   EFFECT OF pH ON IONOGENIC SUBSTANCES

Solute ionization has a pronounced effect on chromatographic behavior, particularly when the pH of the mobile phase is the solute pK_a or pK_b

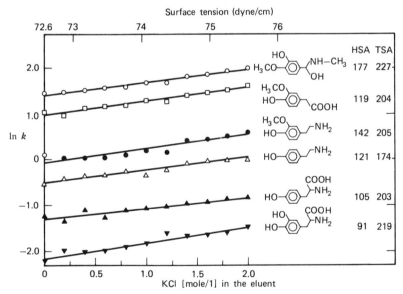

Figure 48   Plot illustrating the effect of salt concentration on the capacity factor; the concentration of KCl in a 0.05 M KH$_2$PO$_4$ solution was varied. The upper scale shows the surface tension of the eluent. The hydrocarbonaceous surface area (HSA) and the total surface area (TSA) of the solute molecules are shown. Column, Partisil 1025 ODS; flow rate, 1.0 ml/min; temperature, 25°C. Reproduced from reference 5 with permission.

value. Very polar compounds and ionic compounds, which are completely ionized in the pH range of 2 to 8, usually give rise to poorly defined and skewed peaks. The peak shape cannot be improved by the adjustment of the composition of mobile phase; however, separation can be achieved on RP packings by using the ion-association technique, which will be discussed later. The retention and peak shapes of weak acids, bases, and zwitterions can also be controlled by ion suppression through the adjustment of the pH of the mobile phase. The extent of this influence can be predicted by using the solvophobic theory and the extended Debye-Hückel theory. In predicting the effect of solute ionization on the capacity factors, it is usually assumed that the rate of equilibration of the ionized and nonionized solute molecules in the mobile phase is considerably higher than the rate of equilibration of the two species with the stationary phase.

5.1 Monoprotic Acids and Bases

The binding of the undissociated solute to the stationary phase and the corresponding equilibrium constants are given by the following relationships:

$$HA + L \rightleftarrows LHA \tag{118}$$

$$K_{LHA} = \frac{[LHA]_s}{[HA]_m[L]_s} \tag{119}$$

where L stands for the ligand, and the subscripts s and m refer to the stationary and mobile phases, respectively (25).

Similarly, the association of the anion, A^-, with the stationary phase can be expressed as follows (25):

$$A^- + L \rightleftarrows LA^- \tag{120}$$

$$K_{LA} = \frac{[LA^-]_s}{[A^-]_m[L]_s} \tag{121}$$

The capacity factor can be expressed in terms of the phase ratio, ϕ, and the distribution ratio of the solute in the stationary and mobile phases (25):

$$k' = \frac{[LHA]_s + [LA^-]_s}{[HA_m + [A^-]_m} \tag{122}$$

Upon substitution of the equilibrium constants K_{LHA} and K_{LA^-} and the acid dissociation constant K_{am} in the mobile phase, the following expression is obtained:

$$k' = \phi \frac{K_{LHA}[L]_s + K_{LA^-}[L]_s \dfrac{K_m}{[H^+]_m}}{1 + \dfrac{K_{am}}{[H^+]_m}} \qquad (123)$$

or

$$k' = \frac{k'_{neut} + k'_{ion} \dfrac{K_{a,m}}{[H^+]_m}}{1 + \dfrac{K_{a,m}}{[H^+]_m}} \qquad (124)$$

where the capacity factor of the neutral solute complex $k'_{neut} = \phi[L]_s K_{LHA}$, and the capacity factor of the ionized solute $k_{ion} = \phi[L]_s K_{LA^-}$ (25). Similarly, the capacity factor for a weak monoprotic base can be derived in an analogous fashion and the following expression is obtained (25):

$$k' = \frac{k'_{neut} + k'_{ion} \dfrac{[H^+]_m}{K_{a,m}}}{1 + \dfrac{[H^+]_m}{K_{a,m}}} \qquad (125)$$

where $K_{a,m}$ is the acid dissociation constant of the protonated base:

$$K_{a,m} = \frac{[H^+]_m [B]_m}{[BH^+]_m} \qquad (126)$$

The effect of eluent pH on the capacity factors of weak acids and bases is shown in Fig. 49.

5.2 Diprotic Acids and Zwitterions

The capacity factors for weak diprotic acids and zwitterions can be calculated by using the appropriate protonic equilibria and association constants (25).

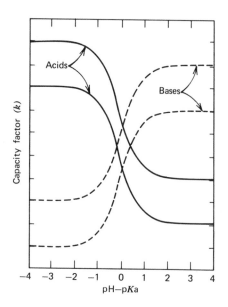

Figure 49 Effect of the eluent pH on retention of weak acids and bases on nonpolar stationary phases. The scale on the abscissa shows the difference between the pH of the eluent and the pK_a value of the solute in the eluent. Reproduced from reference 16 with permission of Preston Publications, Inc.

The k' for diprotic acids can be given as:

$$k' = \frac{k'_{neut} + k'_{ion_1}\dfrac{K_{a_1,m}}{[H^+]_m} + k'_{ion_2}\dfrac{K_{a_1,m}K_{a_2,m}}{[H^+]_m^2}}{1 + \dfrac{K_{a_1,m}}{[H^+]_m} + \dfrac{[H^+]_m}{K_{a_2,m}}} \tag{127}$$

where k'_{neut}, K'_{anion}, and k'_{cation} are the capacity factors of the zwitterionic, anionic, and cationic forces of the solute, respectively (25). All other symbols have the same meaning.

6 EFFECT OF TEMPERATURE

Contrary to the situation in GC, the role of temperature is usually neglected in LC. This is mainly due to the low boiling points of most of the commonly used solvents in RPLC, which limits the operating temperature to 15 to 100°C. In addition, great control of capacity factors can be achieved in RPLC by altering the composition of the mobile phase, so that solvent programming is used instead of temperature programming.

However, the importance of temperature in RP separations should not be underestimated, since in addition to the change in capacity factors, increased column temperature will also increase the overall efficiency of the chromatographic system.

The dependence of the capacity factor on temperature is described by the following relationship:

$$\log k' = \frac{\Delta H^\circ}{2.3RT} + \frac{\Delta S^\circ}{2.3RT} + \log \phi \qquad (128)$$

where ΔH° is the enthalpy of transfer from the stationary to the mobile phase, ΔS° is the accompanying change in entropy where the concentration of the solute in both phases is in molar units, and ϕ is the phase ratio. Generally, a 10° increase in temperature decreases retention by a factor of 2 to 3 (39,40).

If it is assumed that ΔH°, ΔS°, and ϕ are independent of temperature, the relationship takes the form of the classical van't Hoffs plots (log k' versus $1/T$), which are used to estimate the enthalpy and entropy of transfer. It should be pointed out that the measurement of ΔS° is difficult, due to uncertainties in the determination of the surface of the adsorbent and thus the phase ratio, ϕ (38). The slopes of the van't Hoffs plots give straight lines both in partition and adsorption, providing the standard enthalpy of transfer remains constant (38). Thus, in LC where the useful temperature range is between 15°C and 80°C, straight lines are obtained. The magnitude of the slope depends on the nature of the solute (39).

An increase in column temperature is also advantageous in reducing the viscosity of the solvents. This is particularly important with water-methanol mixtures (maximum viscosity at 30% methanol, 1.4 cP), where increased temperature leads to a decrease in the pressure drop necessary to maintain a desired flow rate (39).

It has been reported that increases in column temperature lead to a concomitant increase in column efficiency (41). However, at a constant solvent flow rate, the improvement in HETP was reported to be solely due to the increase in solute diffusivity in the mobile phase, D_m, with decreased solvent viscosity, since the reduced HETP is independent of temperature. It has also been observed that the column capacity increased with increasing temperature (40). The increased capacity is probably due to the increased solute solubility in the mobile phase (39). Changes in temperature may also cause a change in relative retention times when different solutes in a given mixture are retained by different mechanisms of interaction. This effect is due to the differences in the heat of transfer between the two phases.

Finally, an increase in temperature results in improvements in peak symmetry. This effect depends on the chemical nature of the solute and results from the linearization of the distribution isotherms (41).

The use of elevated temperature is particularly advantageous when polymeric phases are employed. The improved peak symmetry may be due to greater linearity of the distribution isotherm at higher temperature or to greater flexibility of the polymer structure (41). The latter effect enables faster mass transfer through the polysiloxane film. This is counteracted by decreased solute permeability in swollen polymeric phases at higher temperatures (41).

The knowledge of the dependence of k' values on temperature is advantageous in the determination of thermodynamic parameters such as the enthalpy of distribution and some parameters affecting entropy (41,42). While the assessment of entropy values is difficult and estimations are plagued with inaccuracies, the enthalpies can be determined relatively easily.

7   EFFECT OF ALKYL CHAIN LENGTH

The effect of chain length of bonded reversed phases in HPLC has received relatively little attention. However, it is generally accepted that greater utilization of the role of n-alkyl groups in reversed phases should be made, since marked changes in retention and loading capacity can be achieved. In general, the structure of the alkyl chain plays a definite role in the control of retention, and possibly the selectivity (1); increased chain length or increased carbon content results in longer retention under given mobile phase composition (1,15,42a,43–46).

It has been stated before that for a given bonded group, the quantity log (k'/S_{BET}) is independent of the specific surface area. Furthermore, when log (k'/S_{BET}) is plotted against the number of carbon atoms in n-alkyl chains, straight lines are obtained (20,44). This indicates that the total area of the n-alkyl chain is accessible for the solvophobic interactions (13,44). Stated differently, the absolute retention of a given solute increases with increasing chain length of the bonded phase. Figure 50 illustrates the dependence of k' on percent carbon of the n-alkyl chain.

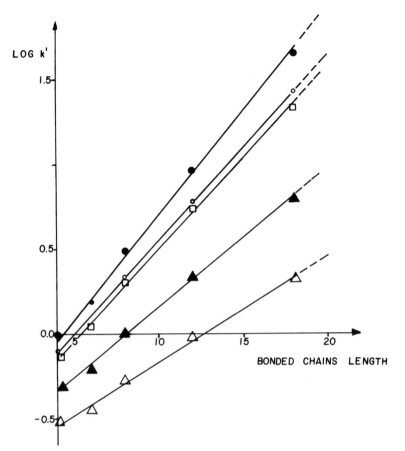

Figure 50 Variation of the logarithm of the capacity factor of aromatic hydrocarbons with the alkyl chain length for a surface coverage of 2.1 μmol/m⁻². Mobile phase, methanol-water (70:30); △, benzene; ▲, naphthalene; □, phenanthrene; ○, anthracene; ●, pyrene. Reproduced from reference 44 with permission.

The slope of the plot increases with increasing chain length and increasing degree of surface coverage (20). However, contradictory opinions exist in the literature. Berensen and de Galan (42a) investigated retention behavior as a function of varying alkyl chain lengths (RP-1 to RP-22) with a methanol-water mobile phase, and contrary to other investigators (1,-15,43,44,46), a continuous increase in retention with increasing alkyl chain length was not observed. The retention was found to increase rapidly up to the "critical chain length" (C_6 to C_{10}), after which it gradually

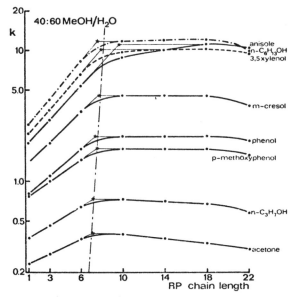

Figure 50a Capacity factors as a function of RP chain length in methanol-water (40:60) and pure water. Asterisks denote the critical chain length, determined at the intersection point of the extrapolated branches of the curves. Reproduced from reference 42a with permission.

leveled off (Fig. 50a). Different solutes require slightly different "critical chain lengths." Stated differently, strongly retained solutes exhibit constant capacity factors with longer alkyl chain length. Weakly retained solutes would require a minimal chain length of six carbons for sufficient retention. Furthermore, since the "critical chain length" for a given solute is independent of the mobile phase composition (42a), these observations suggest that only a certain portion of the alkyl chain length takes part in the separation (42a). The efficiencies of columns packed with phases of different chain lengths are not affected by the length of the bristles (42a,44). However, with phases containing an alkyl chain of less than four carbon atoms, lower efficiencies were observed (44). This is believed to be due to the participation of unreacted silanol groups in the separation mechanism (43).

The effect of chain length on selectivity, α, which is a measure of the thermodynamic difference in the distribution of solutes between the two phases, is a controversial subject. Some authors have found that the selectivity, while dependent on the molecular structure of the solutes and the mobile phase composition, is relatively independent of the chain

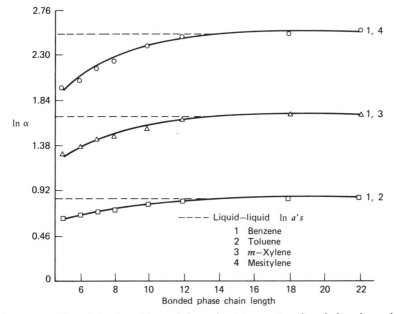

Figure 51 Plot of the logarithm of the selectivity against bonded carbon chain length of the alkyl phase. The dashed lines give the ln α values obtained for the liquid-liquid (nonbonded) systems. Reproduced from reference 12 with permission of Preston Publications Inc.

length (20,45). Others have observed an initial increase in selectivity (C_6 to C_{14}) with increasing chain length, after which it remains constant (12,42a,43,46). Since the average distance between the bonded alkyl chains is approximately 9.0 to 9.5 Å, considerable overlay of the ligands may occur with long alkyl chains (1). Thus it is possible, at least for small solute molecules, to be intercalated in "liquid-droplet clusters" (12), rather than adsorbed on the bonded phase. This would mean that the plateau region in the plot of ln α versus chain length (Fig. 51) is due to the existence of a liquidlike behavior which culminates at C-12 (12). However, the driving force for the retention of solutes would still be the reduction of the total solvophobic area of the solute. With larger solute molecules, the plot of ln α versus chain length (Fig. 52) does not exhibit a plateau region within the same range (12). This is believed to be due to incomplete interaction between the large solute surface and the alkyl chain (12).

The column loading capacity can be determined by measuring the increase in h values and the decrease in k' with increasing sample size at constant flow rate (15). It has been found that the bonding capacity

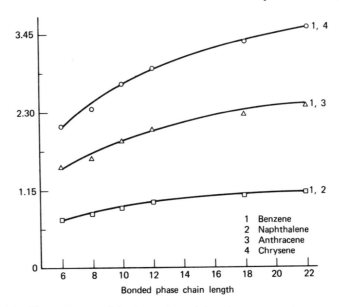

Figure 52 The variation of the logarithm of the chromatographic selectivity as a function of bonded alkyl chain length for a family of larger hydrocarbonaceous solutes. Reproduced from reference 12 with permission of Preston Publications Inc.

increases with increasing chain length (twofold from C_4 to C_{18}) (15). In addition, the maximum sample size parallels the increase in carbon content, but the relationship is not linear (15). The general concensus is that separations of nonpolar and moderately polar compounds on RP columns are best carried out on phases with long alkyl chains such as octadecyl. The use of long alkyl chains gives rise to long retention, which can always be optimized by increasing the amount of organic modifier in the mobile phase (15).

REFERENCES

1. H. Hemetsberger, P. Behrensmeyer, J. Jenning, and H. Ricken, *Chromatographia*, **12**, 71 (1979).

2. D. C. Locke, *J. Chromatogr. Sci.*, **12**, 433 (1974).

3. J. J. Kirkland and J. J. DeStefano, *J. Chromatogr. Sci.*, **8**, 309 (1970).

4. J. A. Knox and A. Pryde, *J. Chromatogr.*, **112**, 171 (1975).

5. C. Horváth, W. Melander, and I. Molnár, *J. Chromatogr.*, **125**, 129 (1976).

6. T. Hanai and K. Fwimura, *J. Chromatogr. Sci.*, **14**, 140 (1976).

7. R. E. Leitch and J. J. DeStefano, *J. Chromatogr. Sci.*, **11**, 105 (1975).

8. M. J. Telepchak, *Chromatographia*, **6**, 234 (1973).

8a. R. P. W. Scott and K. Kucera, *J. Chromatogr.*, **142**, 213 (1977).

9. A. Pryde, *J. Chromatogr. Sci.*, **12**, 486 (1974).

10. V. Rehak and E. Smolkova, *Chromatographia*, **9**, 219 (1976).

11. D. C. Locke, J. J. Schermud, and B. Banner, *Anal. Chem.*, **44**, 90 (1972).

12. C. H. Löchmuller and D. R. Wilder, *J. Chromatogr. Sci.*, **17**, 574 (1979).

13. H. Colin and G. Guiochon, *J. Chromatogr., Chromatogr. Rev.*, **141**, 289 (1977).

14. H. Colin and G. Guiochon, *J. Chromatogr.*, **153**, 183 (1978).

15. K. Karch, I. Sebastian, and I. Halász, *J. Chromatogr.*, **122**, 3 (1976).

16. C. Horváth and W. Melander, *J. Chromatogr. Sci.*, **15**, 393 (1977).

17. H. S. Frank and M. W. Evans, *J. Chem. Phys.*, **13**, 507 (1945).

18. D. C. Locke, *J. Chromatogr. Sci.*, **11**, 120 (1973).

19. J. N. Seiber, *J. Chromatogr.*, **94**, 151 (1974).

20. H. Hemetsberger, W. Maasfeld, and H. Ricken, *Chromatographia*, **9**, 303 (1976).

21. H. Hemetsberger, M. Kellerman, and H. Ricken, *Chromatographia*, **10**, 726 (1977).

22. B. L. Karger, J. R. Gant, A. Hartkopf, and P. H. Weiner, *J. Chromatogr.*, **128**, 65 (1976).

23. T. Halicioğlu and O. Sinanoğlu, *Ann. N.Y. Acad. Sci.*, **158**, 308 (1969).

24. O. Sinanoğlu and S. Abdulnur, *Fed. Proc.*, **243**(part III), 12 (1965).

25. C. Horváth, W. Melander, and I. Molnar, *Anal. Chem.*, **49**, 142 (1977).

26. C. Horváth and W. Melander, *Am. Lab.*, **10**, 17 (1978).

27. W. Kauzmann, *Adv. Prot. Chem.*, **14**, 1 (1959).

28. L. B. Kier, L. H. Hall, W. J. Murray, and M. Randic, *J. Pharm. Sci.*, **64**, 1971 (1975).

29. L. H. Hall, B. L. Kier, and W. J. Murray, *J. Pharm. Sci.*, **64**, 1974 (1975).

30. W. J. Murray, L. H. Hall, and L. B. Kier, *J. Pharm. Sci.*, **64**, 1978 (1975).

31. S. R. Abbot, J. R. Berg, P. Achener, and R. L. Stevenson, *J. Chromatogr.*, **126**, 241 (1973).

32. M. H. Lietzke, R. N. Stoughton, and R. M. Fuoss, *Proc. Natl. Acad. Sci. U.S.*, **50**, 39 (1968)

33. H. B. Bull, *An Introduction to Physical Biochemistry*, F. A. Davis, Co., Philadelphia, 1964, p. 80.

34. J. Timmermans, *The Physico-Chemical Constants of Binary Systems in Concentrated Solutions*, Vol. 4, Wiley-Interscience, 1960, pp. 66, 165.

35. C. Carr and J. A. Riddick, *Ind. Eng. Chem.*, **43**, 692 (1951).

36. G. Akerlof, *J. Am. Chem. Soc.*, **54**, 4125 (1932).

37. G. Douheret and M. Morenas, *Compt. Rend.*, **43C**, 729 (1967).

38. H. Colin and G. Guiochon, *J. Chromatogr.*, **158**, 183 (1978).

39. R. E. Majors, *Analusis*, **10**, 549 (1975).

40. J. A. Schmitt, R. A. Henry, R. C. Williams, and J. F. Dickman, *J. Chromatogr. Sci.*, **9**, 645 (1971).

41. J. H. Knox and G. Vasvari, *J. Chromatogr.*, **83**, 181 (1973).

42. H. Colin, J. C. Diez-Masa, G. Guiochon, T. Czajkowska, and I. Miedziak, *J. Chromatogr.*, **167**, 41 (1978).

42a. G. E. Berendsen and L. de Galan, *J. Chromatogr.*, **196**, 21 (1980).

43. R. E. Majors and M. J. Hopper, *J. Chromatogr. Sci.*, **12**, 767 (1976).

44. M. C. Hennion, C. Picard, and M. Caude, *J. Chromatogr.*, **166**, 21 (1978).

45. K. K. Unger, N. Becker, and P. Roumeliotis, *J. Chromatogr.*, **125**, 115 (1976).

46. K. Karch, I. Sebastian, I. Halász, and H. Engelhardt, *J. Chromatogr.*, **122**, 171 (1976).

BIBLIOGRAPHY

Colin, H., N. Ward, and G. Guiochon, *J. Chromatogr.*, **149**, 169 (1978).

Colin, H., and G. Guiochon, *J. Chromatogr.*, **158**, 183 (1978).

VI Ion-Association (Ion-Pair) Technique

Analyses of ionizable compounds often pose problems with respect to solute retention, peak symmetry, and separation efficiency. If adsorption chromatography with polar eluents is used, excessive peak tailing may result, due to the inhomogeneity of the adsorbent surface. In addition, the peaks may be asymmetrical if the degree of solute ionization changes within the chromatographic band.

Highly polar and ionic compounds cannot be analyzed directly using the RP technique since they elute near the void volume. Therefore, ion exchange has been classically used as a separation technique. However, ion-exchange chromatography has drawbacks. First, simultaneous analysis of both ionized and nonionized solutes cannot be carried out with ion exchange. In addition, ion-exchange columns have lower efficiencies and shorter column life times than the RP packings. A reversed form of the conventional ion-pair chromatography, first used by Horváth and Lipsky in 1966 (1), exploits drastic changes in retention of highly polar and ionic compounds in the presence of ion-association (ion-pair) agents that have a substantial hydrophobic moiety. In 1975 Wahlund (2) applied this technique to the RPLC systems, and since then its applications have continued to grow.

In its use over a period of time, this technique has been known as "soap chromatography" (3), "ion-pair chromatography" (4), "paired-ion chromatography" (5), "solvent-generated (dynamic) ion-exchange chromatography" (6,7), "hataeric chromatography" (8), "detergent-based cation-exchange" (6), "solvophobic-ion chromatography" (9), "surfactant chromatography" (10), and more recently, "ion-association chromatography" (11). The multitude of names under which this extremely successful technique is known suggests the confusion concerning the nature of retention. We have adopted the last term—"ion-association chromatography"—which is broad enough to allow even for "mixed" models of retention.

1 THERMODYNAMIC EQUILIBRIA

Basically, there can be four types of equilibria. In the first one it is postulated that the hydrophobic alkyl ions (counter ions, ion-association reagent) are adsorbed onto the RP packing material, which then acts as a dynamic ion exchanger (6,12,13,15). Bidlingmeyer et al. (11) postulate that the ion-association reagent is strongly adsorbed by the nonpolar stationary phase, forming a charged primary layer that is electrostatically compensated for by the eluent counter ions. Solutes with the same charge as the primary layer will experience electrostatic repulsion by the primary layer and will be excluded rapidly. Solutes of opposite charge will be retained by the primary layer and their retention will increase with increasing charge of the primary layer (i.e., increasing concentration of the counter ion). The retention is thus caused by ionogenic (coulombic) interactions of the solute ions with the modified stationary phase.

Since the two limiting equilibria do not always describe the "Physical Reality" (8), in the third model attempts have been made to combine the first two models in order to express the effect of the concentration of the ion-association agent on retention factors.

Horváth et al. (8) propose the fourth equilibrium, the dynamic complex exchange. According to this model, ion pairs formed in the mobile phase bind to the stationary phase covered with the ion-association agent. Retention takes place by a subsequent metathetical process (8).

According to the second model, ion pairs are formed between the hydrophobic alkyl chain and the charged solute in the mobile phase, and retention is due to the interaction between the neutral pair and the nonpolar stationary phase (2,4,8). Westerlund and Theodorsen (14) suggested that ion-pair formation in the mobile phase and adsorption of the free ion take place.

Proposed equilibria involved in RP ion association are illustrated in Table 21. Although the precise mechanism is not yet clearly established (16), intuitively, it is to be expected that the mechanism of interaction will change with increasing size of the counter ion. Thus with an increasing number of methylene groups in the alkyl chain, the mechanism may shift from ion association to ion exchange. However, it cannot be overemphasized that these equilibria represent limiting cases "and the retention process is not expected to follow any of them over a wide range of chromatographic conditions" (8).

Table 21   Equilibria Involved in RP Ion-Pair Systems as Proposed in the Literature[a]

	A	B	C	D
mobile phase	RNH_3^+	$RNH_3^+ + S^-$	RNH_3^+	$RNH_3^+ \quad Na^+$
	$\uparrow\downarrow$	$\uparrow\downarrow$	$+$	$+ \qquad +$
	RNH_3^+	RNH_3S	$S^- \rightleftarrows RNH_3S$	$NaS \rightleftarrows RNH_3S$
stationary phase				

[a](a), Adsorption of noncomplexed sample ions; (b), complexation in the mobile phase followed by adsorption of the neutral ion pair; (c), complexation on the stationary phase with previously adsorbed amphiphilic ions; (d), ion exchange with buffer ions on adsorbed amphiphilic ions. Reproduced from reference 15 with permission of the publisher.

Recent investigations (11,12) point out that with long-chain counter ions, the ion association does not take place in the mobile phase. This hypothesis is supported by the following findings:

1   A large number of column volumes must be displaced for the ion-association reagent to break through.
2   The capacity factors of the solutes are decreased with mobile phases containing ion-association reagents of the same charge.
3   Conductance measurements demonstrate that ion-pair formation does not occur in the mobile phase.
4   Retention behavior of neutral molecules is independent of the con-centration of ion-association reagent within a certain range.
5   The degree of retention is directly proportional to the surface charge density which arises from adsorption of the ion-association agent.

Regardless of the nature of retention, it can be shown that for the reaction

$$R_{aq}^+ + I_{aq}^- \rightleftarrows [RI]_{org} \tag{129}$$

The following relationship holds true:

$$t_R = \frac{L}{u}(1 + K_{IA}\phi[I_{aq}^-])$$
(130)

where K_{IA} is the overall equilibrium constant for the ion-association process, I_{aq}^- is the concentration of the hydrophobic counter ion, and all other symbols have their usual meaning. It is evident that retention will be governed by the magnitude of the binding constant and the concentration of the counter ion. Additional factors that play an important role in the separation can be summarized as follows:

1 Type of counter ion and degree of interaction with solute.
2 Size of counter ion: increased size leads to longer retention.
3 Concentration of the counter ion: increased concentration results in increased k' values.
4 pH of medium: retention increases if the pH is that at which the solute is completely ionized.
5 Type and concentration of organic modifier in the mobile phase: retention decreases with increasing concentration of the organic modifier.
6 Temperature: temperature has a larger effect in ion association than in other LC techniques, and it is an important variable for optimizing selectivity.
7 Percent coverage of the stationary phase: maximum coverage of the silica surface is required if band tailing is to be avoided.

2 CHOICE OF EXPERIMENTAL PARAMETERS

A judicious choice of the counter ion is a straightforward but important step in setting up an analysis. Most counter ions generally employed are summarized in Table 22.

For the separation of acidic solutes one should start with quaternary amines, and for basic solutes, with alkyl sulfonates. The size of the counter ion has a profound effect on its hydrophobicity and thus the chromatographic behavior of the "paired" species. On the other hand, selectivity may or may not be affected by differences in the size of the

Table 22 Selection of Counter Ions[a]

Type of counter-ion	Main Applications
quaternary amines (tetramethyl, tetrabutyl, palmityl, trimethylammonium ions)	strong and weak acids, sulfonated dyes, carboxylic acids, hydrocortisone and salts
bis-(2-ethylhexyl)phosphate	phenols
tertiary amines (trioctylamine)	sulfonates, carboxylic acids
alkyl sulfonates (methane, pentane-, hexane- or heptane-sulfonate	strong and weak bases, benzalkonium salts, catecholamines, peptides, opium alkaloids, niacin, niacinamide
perchloric acid	forms strong pairs with a wide range of basic compounds e.g. amines, thyroidal iodoamino acids, peptides
alkyl sulfates (octyl-, decyl-, dodecyl-)	similar to sulfonic acid; yields different selectivities

[a]Reproduced (with modification) from reference 16 with permission.

counter ion. It should be pointed out that the size effect can sometimes overshadow the chromatographic behavior of the "paired" species. This is often encountered with the separation of closely related solutes with subtle structural differences, where the use of relatively small counter ions is often advantageous.

Since mixed solvents are often used to separate complex mixtures, care must be taken to ensure that the counter ions are soluble in the least polar solvent (organic modifier). Usually, water-methanol mixtures are employed in reversed phase and most counter ions are readily soluble in methanol. However, in acetonitrile, which may be a preferred solvent due to its lower viscosity, the solubility of most counter ions is considerably lower.

The concentration of the counter ion is an important factor in controlling the retention. A 0.01-F concentration of smaller counter ions and 0.005 F of larger counter ions (dodecyl or larger) will usually suffice (16). Increased concentration of the counter ion beyond a given range may result in a decrease in the k' values (3), probably due to the formation of aggregates or clusters of increased solubility in the mobile phase (3).

In addition, the type of stationary phase and the ionic strength of the eluent may influence solute retention (Bidlingmeyer, 17).

Experimental results show that the relationship between the capacity factor and the counter-ion concentration can be parabolic or hyperbolic.

Horváth et al. (8) have used the solvophobic theory to develop a phenomenological treatment for the determination of the relationship between the capacity factors and various equilibria involved in the chromatographic retention behavior. A particularly interesting parameter, introduced by the same authors, is "the enhancement factor," η, defined by the following relationship:

$$\eta = \frac{B}{k'_0{}^4}P \tag{131}$$

where B is a constant dependent on the mechanism responsible for solute retention in the presence of the counter ion, k'_0 is the capacity factor of the solute in the absence of the counter ion, and P is the stability constant for the binding of the counter ion to the stationary phase. The enhancement factor is practically independent of the solute properties, and it can be used in selecting an optimal counter ion for a given solute or for predicting the effect of the alkyl chain length of the counter ion on the capacity factors (8).

It has been stated before that the selection of the eluent pH is critical in ensuring complete ionization of the solute molecules and maximal ion association. Since most counter ions are commercially available as neutral

Table 23   General Guidelines for Selection of Eluent pH[a]

Type of Solute	Eluent pH	Comment
strong acids (pKa<2)	2–7.4	When solutes are ionized throughout the pH range; pH should be chosen according to other types of solutes present
weak acids (pKa>2)	6–7.4	solutes ionized
	2–5	ionization suppressed; retention influenced by differences in molecular structure of the solute molecules
strong bases (pKa>8)	2–8	solutes ionized throughout the pH range
weak bases	2–5	ionization complete
	6–7.4	ionization suppressed, retention influenced by differences in molecular structure of the solute molecules

[a]Reproduced (with modification) from reference 16 with permission of Preston Publications In

salts, solvents must be buffered to obtain a desired pH value. In determining optimal pH of the eluent, it should be borne in mind that silica-based RP columns are stable in the pH range of 2 to 7.5. In addition, certain buffer ions may also be "paired" with the counter ions and therefore the buffer concentration should be relatively low (0.001 to 0.005 F). General guidelines for the selection of eluent pH are given in Table 23.

REFERENCES

1. C. G. Horváth and S. R. Lipsky, *Nature,* **211,** 748 (1966).
2. K. G. Wahlund, *J. Chromatogr.,* **115,** 411 (1975).
3. J. H. Knox and G. R. Laird, *J. Chromatogr.,* **122,** 17 (1976).
4. B. Fransson, K.-G. Wahlund, I. M. Johansson, and G. Schill, *J. Chromatogr.,* **125,** 327 (1976).
5. *Paired-Ion Chromatography, an Alternate to Ion Exchange,* Waters Assoc., Milford, MA, December 1975.
6. J. C. Kraak, K. M. Jonker, and J. F. K. Huber, *J. Chromatogr.,* **142,** 671 (1977).
7. C. P. Terweij-Groen, S. Heemstra, and J. C. Kraak, *J. Chromatogr.,* **161,** 69 (1978).
8. C. Horváth, W. Melander, I. Molnár, and P. Molnár, *Anal. Chem.,* **49,** 2295 (1977); W. R. Melander and C. Horváth, *J. Chromatogr.,* **201,** 211 (1980).
9. N. E. Hoffman and J. C. Liao, *Anal. Chem.,* **49,** 2231 (1977).
10. E. Tomlinson, T. M. Jeffries, and C. M. Riley, *J. Chromatogr.,* **159,** 315 (1978).
11. B. A. Bidlingmeyer, S. N. Deming, W. P. Price, Jr., B. Sachok, and M. Petrusek, in *Advances in Chromatography,* Proceedings of the 14th International Symposium, Lausanne, September 22–28, 1979.
12. J. H. Knox and J. Jurand, *J. Chromatogr.,* **125,** 89 (1976); J. H. Knox and R. A. Hartwick, *J. Chromatogr.,* **204,** 3 (1981).
13. P. T. Kissinger, *Anal. Chem.,* **49,** 877 (1977).
14. D. Westerlund and A. Theodorsen, *J. Chromatogr.,* **144,** 27 (1977).
15. A. P. Konijnendijk and J. L. M. Van de Venne, in *Advances in Chromatography,* Proceedings of the 14 th International Symposium, Lausanne, September 22–28, 1979.
16. R. Gloor and E. L. Johnson, *J. Chromatogr. Sci.,* **15,** 413 (1977).
17. Bidlingmeyer, B. A., *J. Chromatogr. Sci.,* **18,** 525 (1980) and references contained therein.

BIBLIOGRAPHY

Schill, G., in *Ion Exchange and Solvent Extraction,* Vol. 6, J. A. Marinsky and Y. Marcus, Eds., Dekker, New York, 1974, Chap. 1.

Schill, G., R. Modin, K. O. Borg, and B.-A. Persson, in *Drug Fate and Metabolism,* E. R. Garrett and J. L. Hirtz, Eds., Dekker, New York, 1977, Chap. 4.

Tomlinson, E., T. M. Jeffries, and C. M. Riley, *J. Chromatogr.,* **159,** 315 (1978).

VII Strategy of Developing an RPLC Analysis

RPLC is still a relatively new technique, and literature is not always available on operating conditions for a specific application. Therefore, in this section we will discuss the practical aspects of chromatographic strategy for developing or optimizing an RPLC separation. Although specific types of biochemical analyses will be used as examples, the general method of approach can be adapted to any other group of compounds in similar matrices.

The first step in developing an RPLC analysis, or any other type of chromatographic analysis, is to *define the problem* and *state the purpose of the analysis*. In order to define the problem, the following questions should be asked:

1   Is the analysis going to be used routinely for a large number of samples? Thus are ease of operation and simplicity of great importance?
2   Is a qualitative and/or quantitative analysis required?
3   Is it necessary to separate all the constituents in the sample or only a small group of constituents?
4   Are the constituents similar in structure or widely diverse?
5   Are the constituents present in similar concentrations, or is one constituent present in a large amount and others only in trace amounts?
6   Can the sample be easily prepared for RPLC analysis?
7   Are there compounds present that may interfere with the analysis of constituents of interest?
8   Can the peaks in the chromatogram be readily identified?

For example, in the analysis of serum nucleosides and bases, the purpose of the analysis was to determine the concentrations of all free nucleosides and bases in a small sample (1,2).* The analysis was to be used routinely for a large number of samples. Since nucleosides and bases are present only in picomole amounts, the operating conditions had to be optimized so that nucleotides and other UV-absorbing low-molecular-weight compounds, such as aromatic amino acids and creatinine, would not interfere.

The next step is a *literature search* to find out if these compounds have been separated using other chromatographic techniques. For example, the conditions used in thin-layer chromatography (TLC) or open column chromatography often can be adapted for HPLC; this serves as a starting point and saves valuable time.

A total RPLC analysis involves the following steps:

1   Sample collection.
2   Sample preparation.
3   Chromatography.
4   Peak identification.
5   Quantification.
6   Data analysis and interpretation of results.

1   GENERAL GUIDELINES FOR SAMPLE PREPARATION AND PRECONCENTRATION

Although each sample requires a different approach with respect to the sample collection and preparation procedure, some general remarks are still pertinent. The first step in the analysis of biological samples usually requires sample filtration. Since RPLC columns employ 5- to 10-μm packing materials, the column inlet is usually protected with a 2-μm frit or screen. This frit can be easily plugged with particulate matter which may be present in the sample. Sample filtration can be performed using membrane-type filters with 0.2- to 0.5-μm pore sizes. Deproteination of biological samples is also necessary if long column lifetime and efficient analyses are to be achieved. Several methods of protein removal can be used: ultrafiltration, precipitation of proteins with strong acid or organic solvents, ammonium sulfate precipitation, denaturation by heat, and so on. Some of these methods have been evaluated and comparative studies

---

*References for this chapter are at the end of Chapter X.

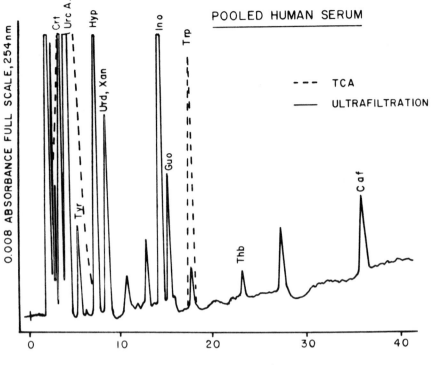

Figure 53 Chromatograms of the UV-absorbing constitutents in normal human serum deproteinated sing the trichloroacetic acid (TCA) and ultrafiltration. Chromatographic conditions are the same as in Fig. 61. Reproduced from reference 1, with permission.

are available in the literature (3,4). It should be pointed out that the deproteination technique can give different recoveries of the compounds of interest. For example, in the study of constituents of pooled human serum, deproteinated by ultrafiltration and with cold trichloroacetic acid (TCA), different levels of the amino acid tryptophan (Trp) were obtained. (see Fig. 53.) The Trp content in the ultrafiltrate represented the *free* amino acid and in the TCA extract, the *total* Trp (i.e., the free Trp and the Trp bound to circulating protein).

Usually, in the analysis of complex samples, only a limited number of compounds are of interest. Therefore, it is not necessary to achieve separation of *all* sample constituents, but rather to optimize the conditions for rapid analysis of *several selected* compounds. If the concentrations of the compounds in a sample matrix differ significantly, the analysis of trace compounds is difficult in the presence of large amounts of interfering species. In these cases, it is advantageous to devise a simple and specific extraction procedure for fractionation of the sample according to the polarity and solubility of the solutes. The efficiency of the extraction procedures must be determined, and ideally, extraction procedures should be highly efficient and specific for the class of compounds of interest.

Extraction procedures can be carried out in different ways. If the procedure is very efficient, only gentle mixing of the sample with the extraction solvent is necessary. Solvents with less favorable partition coefficients may require vigorous shaking, ultrasonic agitation, or exhaustive extraction with a single solvent or multiple solvents. Recoveries can be maximized by careful choice of the solvent polarity (P' factor) for the pH of the sample and/or solvent. If the pH is adjusted so that the compounds of interest are in their ionic form, the extraction solvent is usually water or an alcohol-water mixture with added acid, base, or buffer salt. Conversely, if the pH is such that the compounds are nonionized, a nonpolar extracting solvent is used.

Extraction recoveries can be enhanced by using successive extractions with small volumes of solvent, or by means of exhaustive extractions. In some instances, the compounds of interest may be thermally labile and thus may be subject to decomposition at temperatures resulting from the heat of mixing or high-speed blending. This requires provisions for cooling the sample during extraction.

If the use of extraction solvents gives rise to emulsions that are hard to separate, the addition of salt (e.g., NaCl) or high-speed centrifugation can facilitate the separation. In addition, phase-separating filter papers with hydrophobic coating can be used since they allow only the organic phase to pass through. If the organic extraction solvent has a low melting point, emulsions can easily be separated by centrifugation of the extraction mixture, followed by cooling on dry ice or liquid nitrogen. The organic solvent can then be decanted, leaving the frozen aqueous phase behind.

If the compounds of interest can be preferentially partitioned into low boiling organic solvents, samples can be easily concentrated, if there are no interfering substances present. This is particularly advantageous in multistage extractions where the final volume is large. Preconcentration can be achieved by reducing the volume either by partial evaporation in a rotary evaporator (vacuum and heat) or by a complete evaporation to dryness, followed by reconstitution with an appropriate, less volatile solvent (usually water). In the latter case, care must be exercised near the end of evaporation since volatile solutes may also evaporate.

In some cases, sample cleanup and preconcentration can be achieved by batch adsorption, followed by elution with a suitable solvent. For example, in the analysis of catecholamines in biological samples, adsorption on alumina is often used to separate this class of compounds from other constituents present in higher concentrations (5). Also, in the analysis of ribonucleosides and nucleotides, samples are preconcentrated on a boronate gel column which is highly specific for *cis*-diols (1). TLC, open column chromatography, and low-pressure LC on larger particle packings can also be used for sample cleanup prior to the HPLC analysis. Low-pressure column separations are preferable, since they have higher separation efficiencies and recoveries.

Recently, disposable microcolumns packed with ion exchange or C_{18} packings of 30- to 60-μm particle size have become available. Thus different classes of compounds can be removed from the sample, which greatly facilitates the subsequent RPLC analysis. If the interfering compounds elute with k' values similar to the compounds of interest, it is helpful to remove them by using disposable precolumns. This is the case with highly polar and ionic compounds which elute near the void volume (t_0) in RPLC separations and thus may obliterate the early emerging, less polar solutes of interest.

The use of refillable precolumns with large particle sizes (approximately 40 μm) is becoming an increasingly popular method of sample cleanup; thus untreated samples can be injected directly, sometimes even without removing the protein. Since the precolumns can be dry packed, the packing can be easily replaced when the precolumn becomes overloaded with the impurities. Specific sample preparation methodology pertaining to special applications of both sample matrices and sample solutes is available in the current literature.

2   CHROMATOGRAPHY

The majority of analyses can now be carried out using RPLC; thus RPLC is the method of choice unless the desired separations cannot be achieved, or unless another mode, such as gel permeation, is clearly indicated. At present, with the commercial availability of good RP columns, more than 80% of all RPLC separations are being carried out using C_{18} as a bonded phase on 5- to 10-μm silica particles.

In RPLC, as in any HPLC separation, there are many parameters that can influence both the resolution of compounds in a mixture and the efficiency of the separation. A combination of the appropriate stationary and mobile phases and mode of elution must be determined to obtain the best resolution. Although most separations are carried out at ambient temperatures, some separations will be improved at elevated temperatures. However, regardless of the temperature used, it must be consistent throughout the analysis if reproducible capacity ratios are to be obtained. Unlike GC, temperature programming is rarely used because of the time involved for thermal equilibration of liquids.

After the stationary and mobile phases are chosen, the optimal flow rate and elution mode must then be determined. Only after these conditions have been decided can the separation be "fine tuned" for the constituents of interest, by optimizing the k' and α values. It should be noted that optimization is a balancing process in which the requirements of time of analysis, resolution, and sensitivity are juggled. For example, if it is critical to keep the time of analysis at a minimum, resolution of some components may be lost. This is illustrated with the analysis of adenosine in the presence of nucleotides, nucleosides, and bases (Figs. 54A and B). However, when rapid analysis of only adenosine was needed, resolution of the other nucleosides and bases in the serum was sacrificed to obtain a 7-min isocratic adenosine analysis (Fig. 54C).

2.1 Stationary Phase

The four important parameters involved in the stationary phase that can be varied in RPLC separations are as follows:

1 Column length and internal diameter.
2 Particle size.
3 Type of bonded phase.
4 Surface coverage.

Only after exhausting the many possibilities of mobile phases or mobile phase combinations should different types of columns be tried. However, for certain research projects, or in cases where satisfactory separations cannot be obtained with these columns, stationary phase parameters can

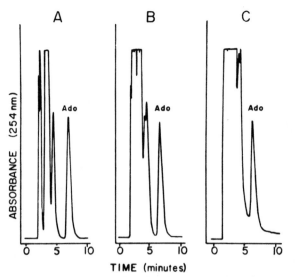

Figure 54 *(A)* Separation of 5 mmol each of seven nucleosides. *(B)* Co-injection of seven nucleosides with 5 mmol each of their bases. *(C)* Separation of Ado from 5 mmol each of the nucleosides, bases, and the mono-, di-, and triphosphate nucleotides: Column, μBondapak/C$_{18}$; eluent, anhydrous methanol—0.007 M KH$_2$PO$_4$ (pH 5.8) (1:9, v/v); flow rate, 2.0 ml/min; temperature, ambient. Reproduced from reference 44 with permission.

be changed singly or in combination. For example, shorter columns (5 to 10 cm) gave better resolution and shorter retention times for the determination of indomethacin and salicyclic acid levels in serum (6). A rapid method is necessary if the analysis is to be used clinically in the treatment of arthritis. Naisch et al. (7) used smaller diameter packings, 5 μm rather than 10 μm, to achieve the desired analysis of theophylline in serum. In an analysis of adriamycin, a phenyl-alkyl bonded phase gave optimal resolution of the parent drug and its metabolites (8), whereas a C$_8$ bonded phase was used to obtain a rapid analysis of short-chain peptide diastereo isomers (9). Thus, in RPLC, various parameters in the stationary phase as well as the mobile phase can be altered to achieve a desired separation.

2.2 Mobile Phase

One of the great advantages of LC is the versatility afforded by a liquid mobile phase. Not only can many different parameters be varied when the mobile phase is liquid, but the solutes can also interact with the mobile phase as well as with the stationary phase. Sufficient solubility of solute molecules in the mobile phase must be ensured in order to prevent precipitation.

For the mobile phase, the first variable to be decided is whether an organic or aqueous eluent should be used. With RPLC analyses, either an aqueous eluent or a very polar organic solvent such as methanol or acetonitrile should be tried first. If the k' values are too large with an aqueous solvent, then one of the organic solvents should be tried. If the k' values are too low with the organic solvent, then separation should be attempted by using a mixture of the two in various proportions. Many simple analyses can be carried out with isocratic elution using an aqueous eluent to which an organic modifier is added. If the sample to be analyzed contains a very complex mixture or a mixture of compounds of diverse structure and retention behavior, then either a ternary mixture of solvents can be used isocratically or gradient elution may be necessary.

Initially, water alone or a very dilute acid, base, or buffer solution is tried. If a buffer is used, the pH as well as the ionic strength of the buffer can be varied. For example, in the separation of nucleosides and bases, it was found that the ionic strength of the mobile phase had little influence on retention behavior, but the pH could affect retention time and/or order of elution (9). The effects of pH could be correlated with pK values; pH had a marked effect on the nucleosides or bases whose pK_a or pK_b values were in the pH range of the eluents, but had no effect on the other purine and pyrimidine compounds (Fig. 55). The buffer cations that can be used include sodium, potassium, or ammonium; the most commonly used anions are phosphate, acetate, and citrate. The stability of the buffer and the ability to maintain the desired pH are very important considerations in choosing a buffering agent.

If the separation cannot be achieved or if the resolution is inadequate, an organic modifier may be added. Isocratic separations with mixed solvents are preferable in the determination of only one or two components in a sample, especially in developing routine clinical assays.

In choosing an organic modifier or eluent, the first considerations are the solubility of the solutes and the compatibility of the solvent with water. Methanol and acetonitrile are usually the organic solvents of choice, although ethanol, dimethyl sulfoxide, dimethyl formamide, tetrahydrofuran, and dioxane are also used. Tables of polarity should be

BASES · NUCLEOSIDES

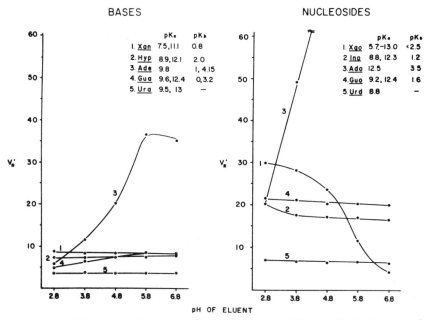

Figure 55 Effect of pH changes on the nucleosides and bases. The ionic strength of the mobile phase was held constant by making all eluents 0.10 F in KCl. Chromatographic conditions: column, μBondapak/C₁₈; low strength eluent, 0.01 M KH₂PO₄ (pH 5.5); high strength eluent, 80:20 (v/v) methanol-water; gradient, linear from 0 to 25% of high concentration eluent in 30 min; flow rate, 1.5 ml/min; temperature, ambient. Reproduced from reference 9 with permission.

consulted to determine the solvent whose polarity is needed to elute the compound of interest in the desired time (Table 2).

More than one organic modifier may be used to achieve the desired resolution, or water may be used as a modifier in a predominantly organic eluent. For example, Koshy and van der Silk (10) used an eluent consisting of CH_3CN, CH_3OH, and H_2O, 94:3:3 by volume, with a 5-μm Zorbax ODS column to separate two metabolites of vitamin D.

A very important and powerful technique for the separation of ionic or ionizable compounds in RPLC is ion association (14,15), discussed in Chapter VI. The use of ion association is illustrated by the work of Hoffman and Liao (11), who separated the highly charged nucleotides using RPLC with tetrabutyl-ammonium ions in the mobile phase (Fig. 56). Previously, these compounds were routinely separated by ion ex-

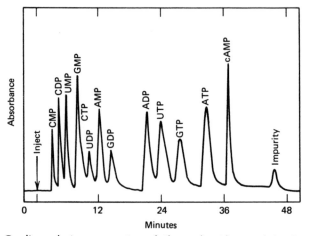

Figure 56 Gradient elution separation of ribonucleotides. Mobile phase: Solvent a, 0.025 M TBHS, 0.050 M KH$_2$PO$_4$, 0.080 M NH$_4$Cl buffer at pH 3.80; Solvent b, 0.025 M TBHS, 0.10 M KH$_2$PO$_4$, 0.20 M NH$_4$Cl buffer at 3.40, 30% methanol. Operating conditions: 40-min gradient program. Reproduced from reference 11 with permission. Copyright (1977) American Chemical Society.

change HPLC (20,21,52,53). However, with the ion-association technique, the nucleosides could also be analyzed along with the nucleotides.

Other reagents may also be used to alter the retention behavior of ionic or ionized compounds. Chow and Grushka (12) used metal ions to improve the k' values of nucleotides. Other compounds such as EDTA have been added to the eluents to improve resolution. For example, Knox et al. (13) used EDTA and an added salt to optimize the separation of tetracyclines.

2.3   Mode of Elution and Flow Rate

Whenever possible, isocratic elution should be used because it eliminates "turnaround time" on the column and thus shortens overall analysis time. Also, retention reproducibility is more predictable with isocratic elution because reequilibration of the column after gradient elution must be carefully controlled. However, when adequate resolution cannot be achieved within a reasonable length of time, because of the diversity of compounds in a mixture or when there is a general elution problem, gradient elution is advisable. Gradients can be either stepwise or continuous. For example, a stepwise gradient was used by Anderson and Murphy (14) to shorten the time of retention and decrease band broadening in the analysis of nucleotides, nucleosides, and bases. Continuous gradients can be of many types: linear, concave, or convex. In addition, the gradient may be delayed and isocratic elution used for the first part of the separation. An example of the power of gradient elution is illustrated in Figs. 57 and 58. All cyclic nucleotides, except cAMP, elute together under isocratic conditions (15), but they are clearly separated using gradient elution (15).

Flow programming is rarely employed. Usually, it does not affect retention order or k' values. However, an increase in flow rate will decrease retention time and peak width. Therefore, if the resolution is adequate and a faster analysis or sharper peaks are desired, flow rate may be increased. Flow rate is related to column length and particle size, as can be seen from the van Deemter equation (Eq. 14). Thus flow rate must be adjusted when column length or particle size is changed and should be optimized for maximum resolution in minimal time.

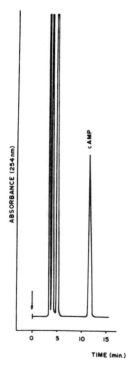

Figure 57 Isocratic elution of reference compounds, optimized for cAMP; chromatographic conditions: column, μBondapak/C_{18}; eluent, anhydrous methanol in 0.02 M KH$_2$PO$_4$ (pH 5.5) (12:88, v/v); flow rate, 1.5 ml/min; temperature, ambient. Reproduced from reference 15 with permission.

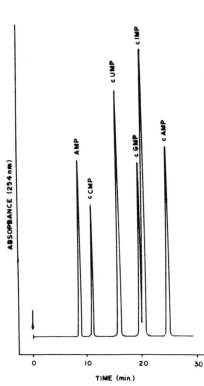

Figure 58 Separation of a synthetic mixture of cyclic ribonucleotides. Chromatographic conditions: column, μBondapak/C$_{18}$; low strength eluent, 0.02 M KH$_2$PO$_4$ (pH 3.7); high strength eluent, methanol-water (3:2, v/v); gradient, linear from 0 to 25% of high strength eluent in 30 min; flow rate, 1.5 ml/min; temperature, ambient. Reproduced from reference 15 with permission.

2.4 Optimizing an RPLC Analysis

Equations are now available for choosing conditions to optimize difficult separations (16–19), especially in gradient elution. They relate separation characteristics to various operating parameters or experimental variables. As yet, these equations are not used routinely in optimization processes and are of little help for unknown samples or for developing new analyses. However, in the future, optimization will undoubtedly be done using mathematical models.

In solving a general elution problem or in "fine tuning" an RPLC analysis, compromise may be necessary among the goals of optimization of analysis time, resolution, and detection sensitivity. Sometimes optimization of one of these parameters is made at the expense of one or more of the other parameters. In this discussion, only changes in the mobile phase will be considered.

To optimize the retention time, many operating parameters should be considered: composition of the eluent, elution mode, and flow rate if gradient elution is being used. To decrease the k' values, the strength of the initial or final eluent or the slope of the gradient should be increased. For example, by increasing the gradient slope from 0.50 to 2.50 in the separation of some nucleosides and bases (19), the analysis time was decreased from 20 to 12 min (Fig. 59). No loss in detector sensitivity was noted.

If isocratic elution does not provide the desired resolution, the most obvious way to improve the resolution is either by isocratic elution with mixed solvents or by gradient elution. The properties of the organic modifier, and the amount added and the type of modifier used in the eluent will affect the resolution. However, if the capacity factors of the compounds to be separated are equally affected by the modifier, then gradient elution may solve the problem. With the gradient, both the slope and gradient time will affect the retention behavior. For example, in the development of a selective analysis for the purine nucleosides and bases, gradient conditions were changed so that the pyrimidines eluted quickly near the void volume and thus did not interfere with the purines (44) Fig. 60).

Optimizing k' and α

Several problems can be encountered in developing an RPLC analysis. These problems can involve favorable k' values but inadequate resolution, poor k' values (i.e., either $k' < 1$ or > 10), or k' values that are very low for one group of compounds and very high for another.

Depending on the types of compounds involved, the problem of inadequate resolution can be tackled by changing the pH of the eluent or the amount or type of organic modifier in the solution. In addition, other modifiers may be used that will alter selectively the retention behavior of one of the solutes. If mobile phase changes cannot provide adequate resolution of the peaks, longer columns (or double or multiple columns) or slightly different types of packings (C_2, C_8, or phenyl instead of C_{18}) can be tried. Another technique that can be used to improve insufficient resolution is "recycling," in which the sample is passed through the column several times until the necessary degree of separation is achieved. Recycling, however, cannot be used with gradient elution.

The optimization of the capacity factors of one compound or a group

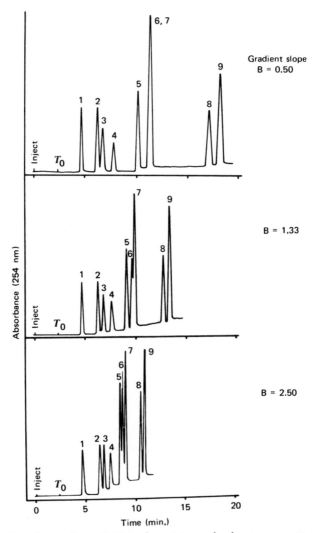

Figure 59 The observed separation of a mixture of reference compounds under three different gradient slopes B. The gradient delay V_p is 3.2 ml, the initial composition C_o is 0% methanol, and the flow rate is 1.0 ml/min. Column, Partisil-5 ODS; low concentration eluent, 0.005 M KH_2PO_4 (pH 5.0); high concentration eluent, methanol-water (3:2, v/v); Peak identity: (1), cytosine; (2), uracil; (3), uric acid; (4), tyrosine; (5), hypoxanthine; (6), uridine; (7), xanthine; (8), inosine; (9), guanosine. Reproduced from reference 19 with permission. Copyright (1979) American Chemical Society.

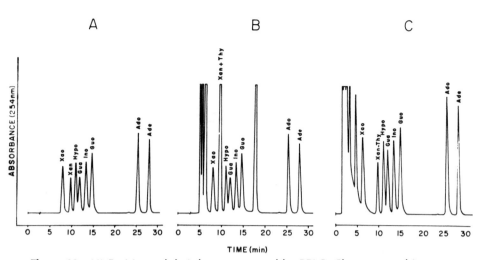

Figure 60   (A) Purines and their bases separated by RPLC. Chromatographic conditions: column, µBondapak/C₁₈; low strength eluent, 0.01 M KH₂PO₄ (pH 5.8);
high strength eluent, 100% methanol; gradient, from 0 to 55% in 30 min; flow rate,
2.0 ml/min; temperature, ambient. (B) Purines and their bases in the presence of
pyrimidines and their nucleosides. (C) Purines and their bases in the presence of
pyrimidines, their nucleosides, and the purine and pyrimidine mono-, di-, and
triphosphaste nucleotides. Reproduced from reference 43 with permission.

of similar compounds usually does not pose many problems. If k' is too
high with an aqueous solvent, the problem may be solved by adding an
appropriate organic modifier until adequate k' values are obtained. If the
capacity ratio is too low and no organic solvent is present, then either
the pH or the ionic strength or composition of the eluent may be changed.
Alternatively, the ion-association technique may be tried. If changes in
the mobile phase do not improve the separation, then increase in column
length or change of packing type may be necessary.

VIII Spectroscopic and Chemical Characterization of Peaks

Although RPLC can now provide excellent resolution of complex mixtures and detect trace amounts of compounds of interest, the elucidation of the chemical nature of the components separated still presents some problems. Since most HPLC detection devices are highly sensitive, making possible the analysis of trace levels of the compounds in small volumes of physiological samples, further characterization of peaks using other methods is often difficult. Therefore, some of the most informative spectroscopic methods such as nuclear magnetic resonance (NMR) and infrared (IR) spectroscopy are not applicable without the Fourier transform facilities (FT) unless the separation is scaled up and a relatively large quantity of pure compounds isolated. The minimum sample size requirements for different, potentially useful characterization methods are listed in Table 24.

Characterization of chromatographic peaks, which can be performed either in the off- or on-line mode, poses certain limitations on the LC system:

1 Mobile phase(s) must be pure (distilled in glass or redistilled solvents).
2 Solvents should be easily evaporated, and the remaining residue should be minimal.
3 Solutes should not react with the mobile phase.
4 Solvents must be compatible with the method by which the solutes are subsequently identified.

Table 24   Minimum Sample Requirements for Some Identification Techniques

	Approximate Sample Size (µg)	
MS	0.005	
Polarography	0.01	
Fluorimetry	0.05	
UV	0.1	
IR (solid)	5	(0.05)[*]
(solution)	50	(0.05)[*]
Spot Analysis	20–100	
NMR	1000	(5)[*]
Elemental Analysis	200–1000	

[*]using Fourier transform methods.

For the off-line identification, LC peaks are usually collected manually, and fraction collectors are needed only for long separations. To avoid cross-contamination of peaks during manual collection, the connecting tubing between the detector outlet and the collection port should be minimal.

Conventional IR spectroscopy is relatively insensitive and restricts the number of useful LC solvents to those with small opaque IR regions. IR spectra can be obtained by careful evaporation of the sample followed by the standard KBr-disc technique or by the attenuated total reflectance (ATR) technique. FTIR can also be performed with solutions using the microsampling technique. Since the on-line FTIR technique employs thin cells (0.1 mm), most common RPLC solvents can be used. Although some spectral regions may be obscured by the absorption of the solvent, functional group information can still be retrieved from portions of the spectrum. In gradient elution, compensation must be made for the continuously changing solvent background by obtaining the spectral background of the gradient run without the sample injection. In addition, new systems are being developed in which solvents are removed prior to the IR analysis using the principle of the wire transport detector (Section 5.6.5).

NMR has similar limitations. However, when coupled with Fourier transform, repetitive NMR scans can be made over an extended period of time and enhanced sensitivity can be achieved. It should also be pointed out that many LC solvents interfere with the NMR analysis, and therefore, they must be removed and the samples dissolved in NMR solvents.

The advantages of mass spectrometry (MS) in providing structural information and molecular weight data were discussed in Section 5.6.10. The off-line use of MS has long been an invaluable tool for the characterization of LC peaks. However, the use of MS is limited by the volatility requirements, and derivatization procedures are usually needed before analyzing thermally labile, nonvolatile, and ionic compounds. Although on-line MS is not currently available for routine use, improvements in direct interfacing of MS-LC will make this technique invaluable for structural elucidation of HPLC solutes.

The use of polarographic analyses, which are very sensitive, for the qualitative analysis in LC is still in its infancy, but it may be valuable for peak characterization in the future.

1   OFF-LINE CHARACTERIZATION

The spot test analyses employing color-formation reactions are relatively insensitive and often necessitate the removal of chromatographic solvents that are incompatible with the reaction medium. However, chemical reactions are available for characterization of functional groups. For example, in the confirmation of the identity of ribose nucleosides and nucleotides (Fig. 61), a simple precolumn periodate (IO_4^-) oxidation may be used. Cleavage of the ribose ring results in altered chromatographic behavior of the reaction product. This reaction is illustrated in Fig. 62. Because of the specificity of the reaction, compounds containing ribose moiety can be identified with a high degree of certainty.

Precolumn enzymatic conversion (enzymatic peak shift) has also been shown to be a simple and elegant method of peak characterization (15,20–22,25,46).* This technique utilizes the specificity of an enzyme in catalyzing a reaction. With a specific enzyme it is possible to prove both the identity and the purity of the substrate and/or product peak. If

---

*References for this chapter are at the end of Chapter X.

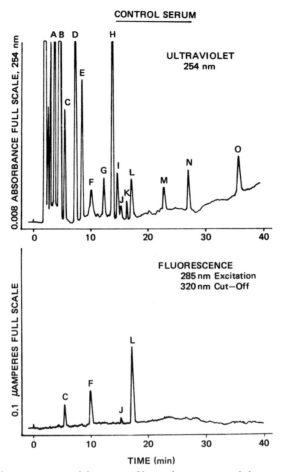

Figure 61 Chromatogram of the serum filtrate from a normal donor, using both UV (254 nm) and fluorescence (285 nm excitation, 320 nm emission) detection. Injection volume, 80 μl; column, μBondapak/C₁₈; low strength eluent, 0.02 M KH₂PO₄ (pH 5.6); high strength eluent, methanol-water (3:2, v/v); gradient, linear from 0 to 100% of the high strength eluent in 87 min; flow rate, 1.5 ml/min; temperature, ambient. Reproduced from reference 1 with permission.

CONTROL SERUM REACTED WITH IO$_4^-$

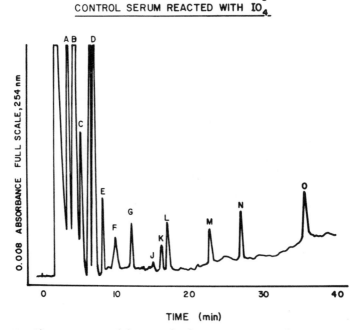

Figure 62 Chromatogram of the sample shown in Fig. 61, after reaction with 80 μl of 0.01 *M* NaIO$_4$ solution. Total sample volume injected, 160 μl. Chromatographic conditions are the same as in Fig. 61. Reproduced from reference 1 with permission.

nonspecific enzymes are used, classes of compounds can be chemically converted to products with different spectral and/or absorption characteristics. The requirements of the procedure include the commercial availability of the enzyme, the ability of the substrate and/or product to respond within the detection system used, and the difference in retention characteristics of the substrate and product. Some of the enzymatic reactions that have been used for characterization of nucleosides and their bases in biological matrices are shown in Table 25. In carrying out enzymatic reactions, an aliquot of the sample is first chromatographed. Another aliquot is buffered to an appropriate pH and incubated with the enzyme. The incubated sample is then chromatographed, and the disappearance of the substrate and/or the appearance of the reaction product confirms the identity and purity of the chromatographic peak.

The use of the enzymatic peak-shift reactions is illustrated with two examples. To confirm the identity of the peak with the retention time of

Table 25 Reactions and Conditions for Some Enzymatic Peak-Shift Reactions[a]

SUBSTRATE(S)	REAGENT OR COFACTOR	ENZYME	REACTION CONDITION pH	PRODUCT
Hypoxanthine	H_2O, O_2	xanthine oxidase (E.C.1.2.3.2)	7.8	xanthine
Xanthine	H_2O, O_2	"	7.8	uric acid
Inosine	phosphate	purine nucleoside phosphorylase (E.C.2.4.2.1)	7.4	hypoxanthine
Guanosine	phosphate	"	7.4	guanine
Guanine	H_2O	guanase (E.C.3.5.4.3)	8.0	xanthine
L-Tryptophan	pyridoxal -5-phosphate	tryptophanase (E.C.4.2.1.E)	8.3	indole
Uric Acid	H_2O, O_2	uricase (E.C.1.7.3.3)	8.5	allantoin
Adenosine	H_2O	adenosine deaminase (E.C.3.5.4.4)	7.5	inosine

[a]Abbreviations: XOD, xanthine oxidase; PNP, purine nucleoside phosphorylase; ADA, adenosine deaminase. Reproduced from reference 29 with permission.

tryptophan in a serum sample, an enzymatic peak-shift reaction with tryptophanase was carried out. The reaction and the conditions are given below:

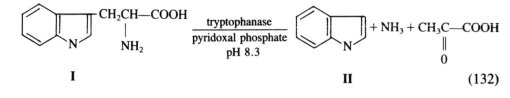

I   II   (132)

Chromatograms of the serum sample and the incubated mixture are given in Fig. 63. Quantitative conversion of tryptophan (I) to indole (II) confirms the peak identity and at the same time "unmasks" the chro-

SERUM — TRYPTOPHANASE   PEAK   SHIFT

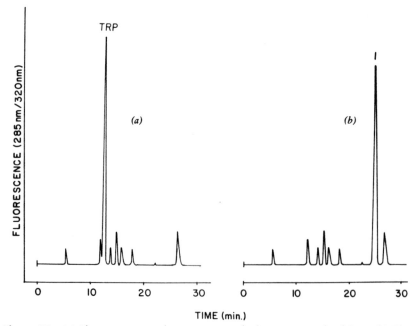

Figure 63 (a) Chromatogram of a serum sample from a normal subject. (b) Chromatogram of the serum sample shown in (a), after incubation with tryptophanase. Peak identity: TRP, tryptophan; I, indole. Chromatographic conditions in (a) and (b): column, µBondapak/C_{18}; low concentration eluent, 0.02 F KH_2PO_4 (pH 5.5); high concentration eluent, methanol-H_2O, (6:4, v/v); gradient, linear from 0 to 40% of high concentration eluent in 35 min; flow rate, 1.5 ml/min; temperature, ambient.

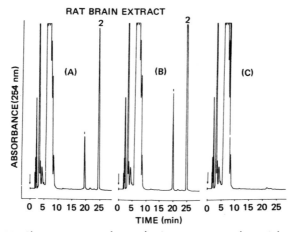

Figure 64 *(A)* Chromatogram of a rat brain extract sample weight corresponding to the volume injected, 3.0 mg; volume injected, 25 μl. Chromatographic conditions: column, μBondapak/C$_{18}$; low strength eluent, 0.02 M KH$_2$PO$_4$ (pH 3.7); high strength eluent, methanol-water (3:2, v/v); gradient, linear from 0 to 25% of the high strength eluent in 30 min; flow rate, 1.5 ml/min; temperature, ambient. *(B)* Chromatogram of the sample shown in *(A)* co-injected with cGMP and cAMP. *(C)* Chromatogram of the sample shown in *(A)* incubated with diesterase (for 10 min). Reproduced from reference 15 with permission.

matogram to prove that no impurities were hidden underneath the tryptophan peak.

A combination of the enzymatic peak-shift reaction and co-chromatography with the reference compounds is illustrated with the RPLC analysis of 3′,5′-cyclic ribonucleotides in rat brain extract. The separation of the constituents of a sample of rat brain tissue is shown in Fig. 64A. Since peaks 1 and 2 had the retention characteristics of adenosine 3′,5′-cyclic phosphate (cAMP) and guanosine 3′,5′-cyclic phosphate (cGMP), an aliquot of the sample was co-chromatographed with the corresponding reference compounds (Fig. 64B). Finally, the extract was incubated with cyclic nucleotide phosphodiesterase (PDE) under the conditions given by the following reaction:

Nucleoside 3′,5′-cyclic phosphate
$$+ \ H_2O \ \xrightarrow[\text{pH 7.5}]{\text{PDE}} \ \text{nucleoside 5′-phosphate} \qquad (133)$$

A chromatogram of the incubated extract is shown in Fig. 64C. Complete removal of peaks 1 and 2, coupled with retention data and UV spectra, confirmed the identity of the peaks.

2 ON-LINE CHARACTERIZATION OF CHROMATOGRAPHIC PEAKS

Since many LC laboratories are not equipped with sophisticated instrumentation which requires skilled personnel, chemists have designed simple, sequential procedures that can be used as an aid in the identification of peaks in routine analysis. Some of these methods are outlined in Table 26. The on-line use of IR spectroscopy has still not found widespread use. Direct MS-LC interface, which provides unambiguous identification of LC solutes, was discussed earlier (Section 5.6.10).

2.1 Retention Behavior and Co-Chromatography with Reference Compounds

Retention behavior of LC solutes (t_R values) is used only for preliminary characterization of chromatographic peaks. Identity assignments are made based on a comparison of retention data for the reference compounds and the peaks in the unknown sample. However, direct comparison of the t_R values necessitates the knowledge of the sample composition and commercial availability of reference compounds.

Retention behavior alone is not sufficient for positive identification, since many compounds can elute with the same retention time. This is

Table 26   Methods for On-line Characterization of Chromatographic Peaks

1)	Retention behavior	
2)	Co-injection with reference compounds	
3)	Using UV-VIS detectors:	a) absorbance ratios b) UV-VIS spectra
4)	Using fluorescence monitors:	a) specificity of native fluorescence b) excitation and/or emission spectra
5)	Using electrochemical detectors:	Ratio of responses at different oxidation or reduction
6)	Using IR detectors: IR spectra	
7)	Chemical derivatization reactions	
8)	LC/MS interface	

especially true in the analyses of complex mixtures of closely related compounds with similar chromatographic properties. For example, peak 1 in the chromatogram of a sample of pooled serum (Fig. 65a) had the retention time of L-kynurenine. When an aliquot of the sample was co-chromatographed with the kynurenine reference solution, the area of the peak under study increased quantitatively (Fig. 65b). However, absorbance ratios at several wavelengths and stopped-flow UV spectra of the peak in pooled serum and the kynurenine reference compound (Fig. 66) indicated that the peak of interest in pooled serum was either not kynurenine or that it contained an unresolved impurity. In some cases, characterization of a peak by direct comparison of its t_R to that of a reference compound may be difficult, due to the sample-matrix effects which can alter the t_R values. Therefore, matching of the bands should be performed using the standard addition technique in which small amounts of reference compounds are added to the sample prior to the chromatography.

It should be pointed out that in spite of the availability of retention data in the literature (see *Journal of Chromatography,* bibliography reviews), these values are at best useful only for rough estimation of t_R values in a nominally equivalent LC system. This is due to the relatively poor column-to-column reproducibility and the fact that retention values are difficult to duplicate even under identical operating conditions. Usually, k' values are more useful for comparison of bands on the basis of retention data, since relative retention is not influenced by solvent flow rate or column geometry. A better method for partial characterization is the chromatographic cross-check which has been employed in gas-liquid chromatography (GLC). This method involves chromatographing of the sample on a different column (if available), where the solute-stationary phase interactions give rise to different retention behavior. For example, if C_{18} columns are used in RPLC, then a different type of RP column (C_2, C_8, or cyano) or normal phase LC should be tried. Alternatively, samples can be analyzed by a different chromatographic technique such as GLC or TLC.

Occasionally, when reference compounds are not available for direct comparison of k' values, retention behavior can be predicted even if the exact mechanism of retention is not precisely known (19,51). In addition to the use of predicted k' values in the tentative characterization of peaks, these values can also be used to determine the solvent strength necessary to elute a compound with the desired k' value, to optimize complex separations, and to predict the structure of a possible internal standard.

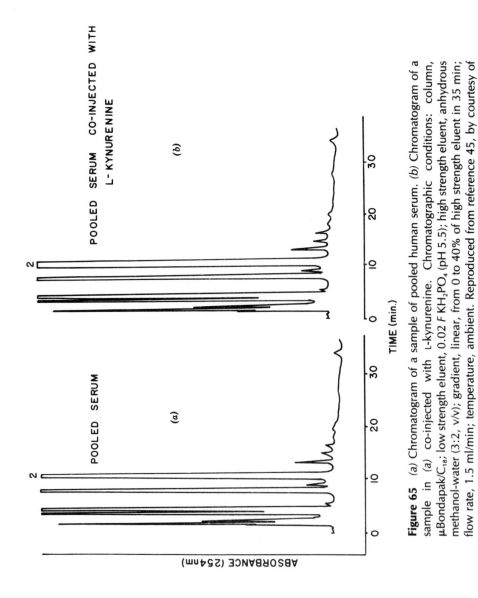

Figure 65 (a) Chromatogram of a sample of pooled human serum. (b) Chromatogram of a sample in (a) co-injected with L-kynurenine. Chromatographic conditions: column, μBondapak/C₁₈; low strength eluent, 0.02 F KH₂PO₄ (pH 5.5); high strength eluent, anhydrous methanol-water (3:2, v/v); gradient, linear, from 0 to 40% of high strength eluent in 35 min; flow rate, 1.5 ml/min; temperature, ambient. Reproduced from reference 45, by courtesy of Marcel Dekker, Inc.

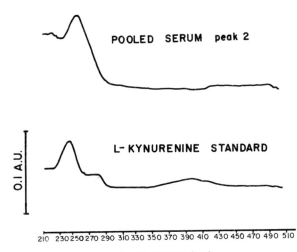

Figure 66 Corrected UV spectra of peak 2 and L-kynurenine reference solution, obtained by stopped-flow scanning. Scanning rate, 100 nm/min; absorbance, 0.4 a.u.f.s. Reproduced from reference 45, by courtesy of Marcel Dekker, Inc.

For structurally similar compounds, a relationship usually exists between some function of t_R or k' value and the number of structural units that are repeated in the sample. Under isocratic conditions, the relationship is given by the Martin rule:

$$\log k' = A + Bn \qquad (132)$$

where A and B are constants for the homologous series under the chromatographic conditions used, and n is the number of repeating groups. The validity of Eq. 132 is illustrated in Fig. 67, which shows a linear relationship between $\log k'$ and the number of aliphatic carbon atoms in the RPLC separation of homologous carboxylic acids.

Similar plots have been obtained by correlating $\log k'$ and $\log P$ (extraction coefficient). The P values have been obtained from equilibrium distribution systems (classical Shake-Flask procedures) (41). The correlation between $\log k'$ and $\log P$, illustrated in Fig. 68, demonstrates the tremendous power of this method in predicting t_R values. Other relationships have been used to predict the retention times and to optimize the

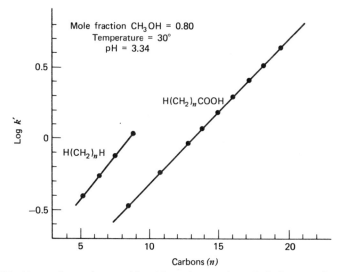

Figure 67 Linear dependence of log k' on the number of aliphatic carbon atoms. Reproduced from reference 40 with permission. Copyright (1977) American Chemical Society.

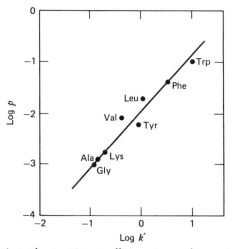

Figure 68 Correlation of extraction coefficient (octanol-water), p, with k' values for some amino acids. Column, Lichrosorb RP-8, (5 μm); mobile phase, 0.1 M phosphate (pH 6.7); temperature, 70°C; flow rate, 2.0 ml/min. Reproduced from reference 41 with permission.

194

separations of different groups of compounds under gradient elution conditions. Several authors have found that for certain classes of compounds, a linear relationship exists between ln k' and the percent of organic modifier in the mobile phase (18,19). However, if the range of mobile phase composition is wide, higher order equations are necessary to predict the elution of compounds (17).

In one study (19), gradient conditions were predicted for the separation of nucleosides and bases using the slope of the graph of ln k' versus percent organic modifier and the y-axis intercept for each compound (Fig. 69). Once these constants have been determined, they can be used to estimate the retention times of the compounds under different gradient and isocratic conditions. Using the empirical constants and several com-

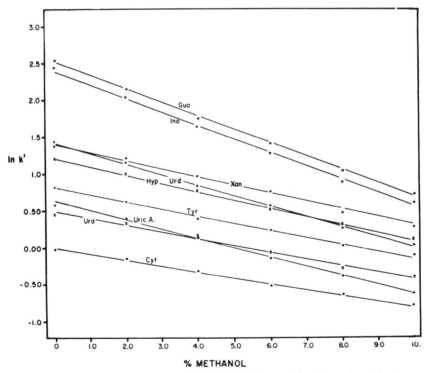

Figure 69 Plot of ln k' versus % methanol in the mobile phase for eight of the nucleosides, bases, and the amino acid tyrosine. All data points represent the average of two injections. Column, Partisil-5 ODS; aqueous portion of the mobile phase, 0.005 M KH$_2$PO$_4$ (pH 5.0); flow rate, 1.0 ml/min; temperature, 25°C. Reproduced from reference 19 with permission. Copyright (1979) American Chemical Society.

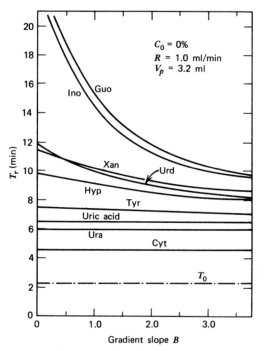

Figure 70 Plot of the predicted retention times versus the gradient slope for the mixture of nucleosides, bases, the amino acid tyrosine, and uric acid. Gradient delay, flow rate, and initial composition were held constant, whereas the retention times were calculated in various gradient slopes B, using a computer program. Chromatographic conditions are the same as in Fig. 69. Reproduced from 19 with permission. Copyright (1979) American Chemical Society.

puter programs, the gradient conditions can be predicted for the separation of several compounds in minimum time with maximum resolution. The relationship between the predicted t_R values and the gradient slope is shown in Fig. 70. The chromatogram in Fig. 59 illustrates the actual separations of nine compounds under the gradient conditions.

2.2   Spectroscopic Characterization of HPLC Peaks

2.2.1   Absorbance Ratios and On-Line UV Scanning

On-line spectroscopic identification methods have been simplified with the introduction of variable wavelength micro-UV detectors. Their wide dynamic range, high sensitivity, and low noise levels have not only enhanced the detectability of trace compounds, but also have given the chromatographer new possibilities for determining peak identities.

The use of absorbance ratios is becoming increasingly popular for characterization of HPLC peaks (22–25). By using two dual wavelength detectors connected in series, it is possible to monitor HPLC effluents at several wavelengths and use the computed absorbance ratios for comparison with those of the reference compounds. Absorbance ratios are particularly helpful in distinguishing between closely eluting compounds. Wavelengths for absorbance ratios should be selected according to the following criteria:

1   Absorption maximum of either the primary compound of interest or the suspected impurity.
2   A general wavelength, such as 254 nm, at which many biologically important compounds absorb.
3   Wavelengths in the 190- to 230-nm range where many compounds exhibit high end absorption.

A combination of absorbance ratios, retention behavior, and co-chromatography was used to characterize peaks in a chromatogram of a serum sample from a patient with congestive heart failure (46). First, a synthetic mixture of selected nucleosides and bases, which are the major low-molecular-weight UV-absorbing constituents in serum, was chromatographed and detected at 254 and 280 nm (Fig. 71). Next, the peaks in the serum sample were tentatively identified on the basis of retention times (Fig. 72). An aliquot of the sample was then co-chromatographed with the reference compounds. Finally, the peak height ratios were computed for the unknown peaks in the serum sample and compared with those of the reference compounds (Table 25). The identity of the chromatographic peaks was deduced from evidence accumulated from retention behavior, co-chromatography with reference compounds, and absorbance ratios.

Although absorbance ratios aid considerably in the characterization of chromatographic peaks, the examination of the entire spectrum is more advantageous than the comparison of absorbances at several points along the spectral curve. In spite of the fact that UV-absorption spectra lack

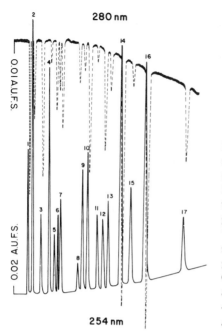

Figure 71 Separation of some nucleosides and bases detected at 254 and 280 nm: 1, Cytosine; 2, uracil; 3, cytidine; 4, hypoxanthine; 5, xanthine; 6, uridine; 7, thymine; 8, xanthosine; 9, inosine; 10, guanosine, 11, adenosine; 12, thymidine; 13, 1 methylinosine; 14, N^2-methylguanosine; 15, adenosine; 16, N^2-N^2-dimethylguanosine; 17, N^6-methyladenosine. Chromatographic conditions are the same as in Fig. 61. Reproduced from reference 46 with permission. Copyright (1977) American Chemical Society.

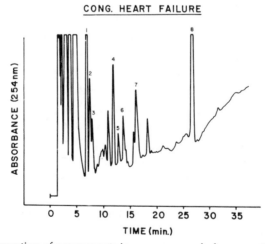

Figure 72 Separation of components in a serum sample from a patient with congestive heart failure. Chromatographic conditions are the same as in Fig. 61. Reproduced from reference 46 with permission. Copyright (1977) American Chemical Society.

fine structure and usually contain broad bands, they are, nevertheless, a fingerprint of the absorber. While the rapid scanning techniques are still not widely used, the stopped-flow UV scanning has proved to be a valuable tool for simple on-line characterization of chromatographic peaks (25). Since the solute diffusivity in the mobile phase is slow, the diffusion effects are not significant, even if the flow is arrested for several hours (25). Usually, provisions are made for obtaining corrected spectra by automatic correction of the spectral background arising from the change in optical properties of the solvent system, flow cells, and monochromator.

The combined use of several characterization techniques is illustrated with the analysis of uric acid and caffeine in normal serum (Fig. 73a). After tentative identification on the basis of co-chromatography with the respective reference solutions (Fig. 73b), the ratios of absorbances were computed for the two unknown peaks detected at 280 and 254 nm and

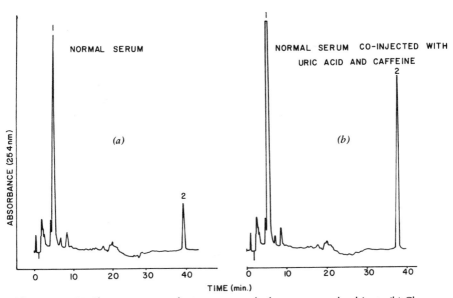

Figure 73 (a) Chromatogram of a serum sample from a normal subject. (b) Chromatogram of a normal sample co-injected with uric acid and caffeine. Chromatographic conditions are the same as in Fig. 61. Reproduced from reference 25 with permission.

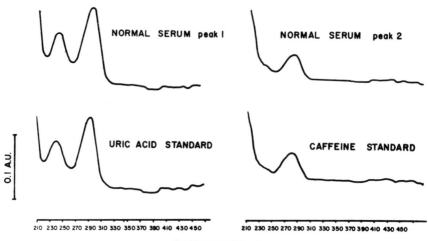

210 230 250 270 290 310 330 350 370 390 410 430 450   210 230 250 270 290 310 330 350 370 390 410 430 450

WAVELENGTH (nm)

Figure 74   Stopped-flow UV spectra of peaks 1 and 2 in normal serum shown in Fig. 73(A) and the uric acid and caffeine reference solutions. Scanning conditions: scanning rate, 100 nm/min. Reproduced from reference 25 with permission.

compared with those of the reference compounds. Close agreement between the ratios gave additional information concerning the identity of the two peaks under study. Next, UV spectra of the two peaks in the serum sample were obtained by the stopped-flow technique and compared with those of the reference compounds (Fig. 74). Combined evidence from all identification steps led to the conclusion that the two peaks in serum were uric acid and caffeine.

2.2.2   Fluorometric Detection

Fluorometric detection offers advantages over UV absorption because of its greater sensitivity and specificity (1,26). Thus it is an ideal tool for trace analysis where interferences must be reduced or eliminated. Since relatively few naturally occurring compounds exhibit native fluorescence, detection is very selective. However, when necessary, compounds that do not fluoresce naturally can be derivatized.

In the analysis of tryptophan (Trp), an essential amino acid, it is desirable to use fluorometric detection to achieve better detectability and to reduce possible interferences. This is illustrated with the analysis of the constituents of a sample of control serum detected both by UV absorption and fluorometry (Fig. 61). The advantages of fluorometric detection can readily be seen in the larger size of the tryptophan peak (peak L) and in the selectivity of the fluorescence detection.

In the determination of catecholamines, compounds of physiological and pharmacological importance, fluorometric detection was mandatory, since UV absorption at either 280 or 254 nm did not provide sufficient sensitivity (47). Figure 75 shows the separation of a synthetic mixture

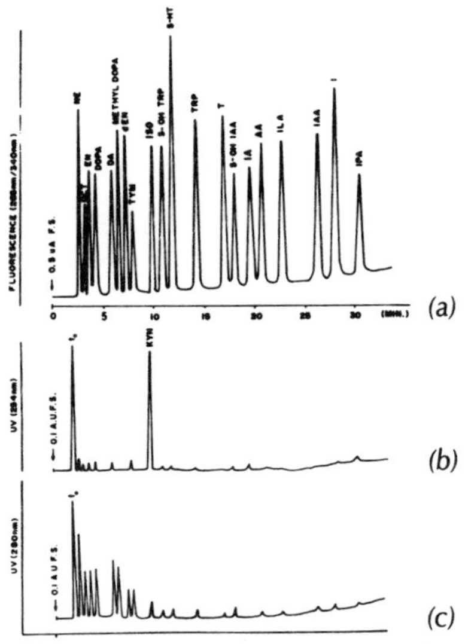

Figure 75   HPLC separation of standard catecholamines and tryptophan metabolites detected by (a) their native fluorescence (285 nm ex., 340 nm, cutoff filter), (b) UV absorption at 254 nm, and (c) UV absorption at 280 nm. Conditions: column, μBondapak/C_{18}; low concentration eluent, 0.02 M KH_2PO_4 (pH 3.7); high concentration eluent, anhydrous methanol-water (6:4, v/v); gradient, linear from 0 to 100% of the high concentration eluent in 35 min: flow rate, 1.5 ml/min, temperature ambient. Peaks: NE, 11.8 nmol; OCT, 18.8 nmol; KYN, 19.8 nmol; EN, 15.9 nmol; DOPA, 9.2 nmol; DA, 20.4 nmol; methyl DOPA, 12.2 nmol; EN, 18.2 nmol; TYM; 17.2 nmol; ISO, 18.5 nmol; 5-OH TRP, 3.9 nmol; 5-HT, 0.92 nmol; TRP, 4.4 nmol; T, 3.9 nmol; 5-OH IAA, 3.1 nmol; AA, 8.2 nmol; IA, 4.0 nmol; ILA, 3.9 nmol; IAA, 4.0 nmol; I, 3.2 nmol; IPA, 4.5 nmol. Reproduced from reference 47 with permission.

of catecholamines and some metabolites detected by measuring their native fluorescence and UV absorption at 280 and 254 nm.

If the fluorometer is also equipped with scanning capabilities, it is possible to obtain stopped-flow excitation and/or emission spectra of the components separated. This provides a powerful means for determining peak identity. The usefulness of the excitation spectra is illustrated by the analysis of catecholamines and related compounds in a sample of rat brain homogenate (47). The separation of the rat brain constituents de-

RAT BRAIN TISSUE

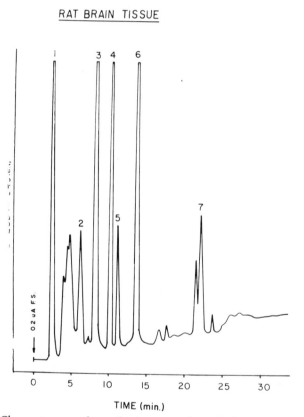

Figure 76 Chromatogram of a rat brain extract (hypothalamus). Chromatographic conditions are the same as in Fig. 75. Reproduced from reference 47 with permission.

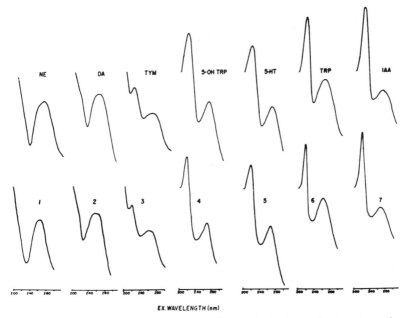

EX. WAVELENGTH (nm)

Figure 77 Corrected excitation spectra of the peaks in the rat brain extract shown in Fig. 76 and the spectra of the corresponding reference compounds. Scanning conditions: scanning rate, 100 nm/min; range, 0.1 a.u.f.s. Reproduced from reference 47 with permission.

tected fluorometrically using an excitation wavelength of 285 nm and emission cutoff filter of 340 nm is shown in Fig. 76.

Initial peak identification, based on comparison of retention times, was confirmed by means of stopped-flow excitation spectra. Figure 77 shows the marked similarity between the excitation spectra of the constituents of rat brain extract and the corresponding reference compounds.

2.3   Chemical Characterization of HPLC Peaks: Electrochemical Detection

The possibility of using hydrodynamic thin-layer electrochemistry for detection of trace amounts of electroactive compounds in HPLC effluents is gaining widespread popularity. Electrochemical detection is highly sensitive, which is particularly advantageous in the analysis of trace components which do not absorb in the UV range or when only limited volumes of physiological samples are available. Compounds can be detected in the oxidative or reductive mode, and the potential applied can be controlled to maximize sensitivity. The increased selectivity of electrochemical detection was found useful for the determination of conju-

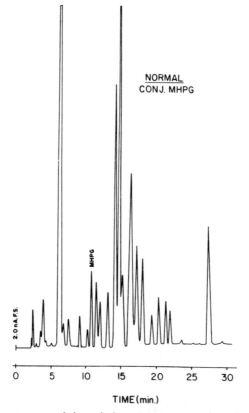

Figure 78 Chromatogram of the ethyl acetate extract of a urine sample from a healthy subject after incubation with glusulase following initial extraction of free acidic compounds. Chromatographic conditions are the same as in Fig. 75: detection, electrochemical at + 1.00 V; volume of extract injected, 5 μl (15 μl urine). Reproduced from reference 48b with permission.

gated MHPG in the urine extract from a healthy subject. The chromatogram of this sample is shown in Fig. 78.

Since each class of compounds exhibits distinctively different electrochemistry at various potentials, these changes in response (current) can be exploited for selective detection and peak characterization. By varying the potential, characteristic changes in the response of the solute will result, and their ratios can be used to characterize a peak. Figure 79 illustrates the change in response of 3-methoxy-4-hydroxyphenylethylene glycol (MHPG) with change in oxidation potential. In addition, hydro-

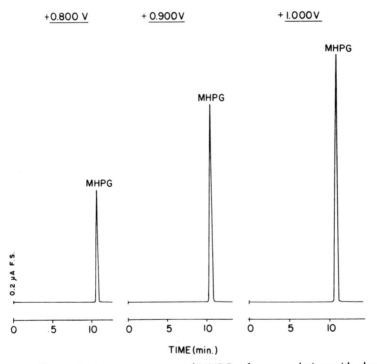

Figure 79 Change in detector response of MHPG reference solution with change in oxidation potential. Chromatographic conditions: column, μBondapak/C$_{18}$; (low strength) eluent, 0.1 F KH$_2$PO$_4$ (pH 2.50); (high strength) eluent, acetonitrile-water (3:2); gradient, linear from 0 to 60% of the high strength eluent in 45 min; flow rate, 1.4 ml/min; temperature, ambient. Reproduced from reference 48 with permission.

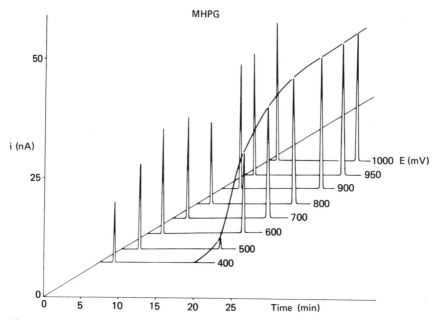

Figure 80   Variation in the MHPG response with change in oxidation potential. Amount injected, 3.0 μg. Chromatographic conditions: column, μBondapak/C$_{18}$; low-strength eluant, 0.1 *M* KH$_2$PO$_4$, pH 2.50; high strength eluant, 3:2 (vol/vol) CH$_3$OH-H$_2$O; gradient, linear, from 0 to 60% of the high strength eluant in 45 min; flow rate, 1.2 mL/min; temperature, ambient; detection, amperometric at + 1.000 v; sensitivity, 2 μA; recorder attenuation, × 1024. Reproduced from Ref. 48b with permission.

dynamic voltammograms can also be constructed in the following manner: repeated injections of the reference compounds and sample are made and the current is recorded at several potentials. If the current at each applied potential is divided by the current at a potential at which the response is maximal, a relative current ratio (φ) is obtained. The plot of φ versus the potential is constructed for the reference compound and the peak of interest in the sample. The comparison of the two voltammograms provides considerable information concerning the peak identity (28) (Fig. 80).

IX Chemical Derivatization

Chemical derivatization as an adjunct to chromatography is by no means a new concept. It has been used in TLC (both for qualitative and quantitative work), as well as in classical UV absorption and fluorescence work, in NMR, MS, and so on. The use of chemical derivatization in LC has several advantages:

1 *To improve extraction recovery of the compounds of interest:* Improved recovery is usually achieved by adjusting the pH of the medium or by means of ion-association extractions, whereby the nonpolar complexed species are partitioned preferentially into the less polar solvent.

2 *To enhance the LC separation:* Chemical derivatization is often used to improve the LC separations of optically active compounds on nonchiral stationary phases. The derivatizing agent is a chiral compound and the reaction product is a diastereomer. Occasionally, derivatization is also used to block hydrogen-bonding sites in order to decrease solute polarity.

3 *To enhance detectability:* Compounds that do not absorb or fluoresce strongly can be derivatized by addition of UV or fluorescent tags. The enhancement of detectability by chemical derivatization, one of the most important advantages of chemical derivatization, is receiving increasing attention.

4 *To confirm peak identity:* Identity of the parent peak can be verified by forming a derivative of known chromatographic behavior.

The importance of derivatization is obvious from the number of derivatization procedures described in the literature. Table 6 lists some general derivatization reagents, along with the reaction products and their spectral properties. Details of the derivatization procedure can be found in the literature. In general, fluorescence derivatization should be carried out with nonfluorescent reagents that yield highly fluorescent products so that excess reagent will not interfere with the separation. Furthermore, reaction by-products should be minimal. UV-tagging reagents most often contain a substituted aromatic chromophore with high molar extinction coefficients (ε = approximately 10^4) (27).* Derivatization can be accomplished either before or after the chromatographic process:

1 Precolumn derivatization:
   (a) Off-line: the reaction is carried out prior to injection of the sample into the chromatograph.
   (b) On-line: the reaction is carried out in the chromatograph between the injector and the column inlet.
2 Postcolumn derivatization:
   (a) Off-line: the reaction is carried out after the chromatographic separation on collected fractions.
   (b) On-line: derivatization is carried out in the chromatograph in a reactor cell placed between the column outlet and the detector.

Precolumn, off-line derivatization is very often used in LC since it does not pose any restrictions on the chromatographic system in terms of the mobile phase composition, reaction temperature and duration, and so on. In addition, sample cleanup can be carried out easily. This is generally not the case with precolumn, on-line derivatization. However, both reactions have the same disadvantage: some derivatization reactions yield several by-products that can complicate the interpretation of the chromatogram. Furthermore, the derivatives have to be stable with no decomposition occurring during their passage through the column.

Postcolumn, on-line derivatization reactions are carried out on separated chromatographic bands. Therefore, the requirements for stability of the derivative and absence of interferences in the separation process are less stringent. However, this type of derivatization places demands on the mobile phase composition, reaction time, temperature, and so on, in addition to the need for a special reactor of small volume to avoid band broadening. Postcolumn, off-line derivatization, which combines the advantages of all other types of derivatization mentioned thusfar, has been extensively used in the past and is reviewed in the literature (27–32).

---

*References for this chapter are at the end of Chapter 10.

X Quantitative Analyses Using RPLC

RPLC has great usefulness and popularity not only because highly complex mixtures can be readily separated, but also because precise quantitative data can be obtained rapidly and reproducibly. Its success in quantitative analysis rivals and often surpasses GLC, particularly in areas where GLC requires elaborate sample pretreatment. High accuracy and precision of data have been made possible due to developments in column technology and improvements in instrument design. Thus trace analysis of parts per billion (ppb) of many compounds can be performed.

It should be pointed out that each step in the quantitative analysis is important and that the overall accuracy of the procedure will be determined by the least accurate step. Many factors can influence the reliability of quantitative data. They can be summarized as follows:

1 Sampling technique, which includes sample collection, cleanup, or preconcentration.
2 Sample size.
3 Efficiency of the chromatographic separation.
4 Accuracy of detection system.
5 Method of quantification.
6 Interpretation of quantitative results.

A full discussion of the sampling technique is beyond the scope of this book and the details can be found in many textbooks of analytical chemistry. Sampling is crucial in all analytical methods. However, its importance is particularly evident in high-resolution techniques for trace analysis, such as RPLC. Suffice it to say that a sample should be homogeneous and that it should represent the bulk material in a reproducible way.

If solutions of solids are to be assayed by RPLC, the material must be completely dissolved, and quantitative dilutions should be carried out using standard, established analytical procedures. For accurate analyses, carefully calibrated volumetric glassware must be used at a controlled temperature. Whenever possible, the solvent used for the dissolution of a sample should be the same as the mobile phase. This is advantageous since it eliminates the possibility of solute precipitation on the chromatographic column. Furthermore, variations in the k' values and band shapes will be minimized. If a similar solvent cannot be found, another solvent may be used, as long as it is completely miscible with the mobile phase.

The method of sample introduction is also important. Generally, the precision of syringe injections is less than that of the valve injectors with calibrated loops. Therefore, the introduction of exceedingly small volumes with syringe injections should be avoided. For best analytical precision, 25- to 100-μl volumes should be injected using the syringe method. The use of large sample volumes has the added advantage of eliminating the need for operating at high detector sensitivity settings, at which baseline noise and drift may interfere with quantification. At the other extreme, concentrations exceeding the column loading capacity should also be avoided in order to obtain reproducible k' values and optimal column plate count.

The efficiency of the chromatographic column is an important factor in quantitative analysis. Complete separation of sample components is desirable since poorly resolved peaks are difficult to quantify even using electronic integrators. Therefore, it is particularly important to ensure that the peaks of interest are eluted with k' values greater than 2 (resolution of early peaks is usually poor) and less than 5 to minimize peak broadening. In addition, peaks should be symmetrical, especially if quantification is performed manually. Microparticulate RP columns with 5- to 10-μm packing materials afford high resolution and efficiency, provided the operating parameters are properly chosen. Therefore, in most instances, the occurrence of peaks with severe tailing is usually indicative of the poor selection of the column or the chromatographic mode.

Since the overall detector sensitivity is a function of flow rate and signal-to-noise ratio, the baseline stability is crucial in trace analysis. Furthermore, the detector response is different for various detection devices and solutes. Therefore, the detector response should be determined for each solute separately over the concentration range of interest.

After all the operational parameters have been optimized and peaks identified unambiguously, peak size can be determined either on the basis of peak height or peak area. As pointed out earlier, the size of a chromatographic peak is proportional to the amount of solute under the peak.

However, the detector can be either mass or concentration sensitive. Most commonly used LC detectors are concentration sensitive, and their response is independent of the rate at which the sample is introduced into the detector. If the composition of the mobile phase is precisely controlled, peak height measurements will be relatively independent of the flow rate. Conversely, if the flow rate can be maintained constant and if the solvent composition changes due to evaporation of volatile solvents or poor reproducibility of gradient programs, peak area is the preferred measurement. Under these conditions, peak heights may be significantly affected by changes in solvent composition which would affect the precision of quantification. Thus flow rate must be carefully controlled if peak areas are measured, and solvent composition (gradient) must be maintained constant for peak height measurements.

1   MEASUREMENT OF PEAK HEIGHT AND AREA

The peak height measurement (h') represents the vertical distance from the peak maximum to the baseline. If peak heights are used for quantification, the baseline should be stable and the noise minimal. In some cases, particularly when the mobile phase composition changes drastically during gradient elution, a baseline drift may result. This drift can be corrected by extrapolating the baseline between the front and the end of the peak tracing. Different methods of measuring peak heights are illustrated in Fig. 81. Peak heights are less affected by incomplete resolution of neighboring peaks than peak areas. However, peak area measurements are mandatory in quantification of asymmetrical peaks. Measurements of

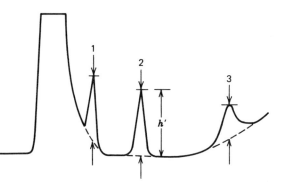

Figure 81   Peak height measurements. Reproduced from reference 42 with permission of Varian Associates, Inc.

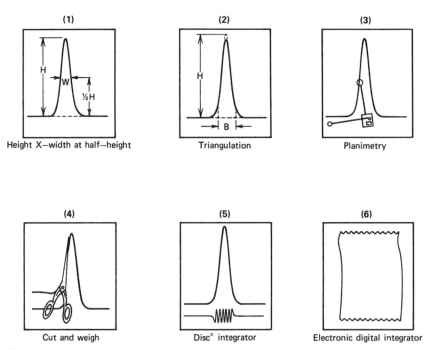

Figure 82 Techniques for measuring peak areas. Reproduced from reference 42 with permission of Varian Associates, Inc.

peak areas can be performed by several methods, which are illustrated schematically in Fig. 82:

1 Peak height multiplied by width at half-height $w_{1/2}$.
2 Peak height multiplied by width at peak base (triangulation).
3 Mechanical integration using a planimeter.
4 Cut-and-weigh method.
5 Mechanical integration using a ball-and-disc integrator.
6 Digital electronic integrators.
7 Computing integrators.

Methods 1 and 2 are self-explanatory and will not be dealt with in great detail. The "height times width at half-height" assumes Gaussian peak shape, and it circumvents the uncertainties resulting from the construction of tangents from peak sides used in triangulation. With both methods, the ratio of peak height to width should not be very large, since significant errors result from measurements of very narrow peaks.

Planimetry is a tedious method of mechanical integration which is now rarely used. The area measurements are performed by tracing the perimeter of the peak, which is time-consuming and requires considerable skill.

The cut-and-weight method, besides being destructive, also requires considerable skill. In addition, weight constancy of the chart paper is mandatory for reproducible and accurate measurements.

With mechanical ball-and-disc integrators, peak perimeter is traced using a mechanical driver. Although the method is well suited for irregularly shaped peaks, incomplete resolution and baseline drifts cannot be taken into account.

More sophisticated digital electronic integrators and computing integrators are a common feature in most laboratories today. These devices possess a great deal of flexibility: they usually take into account baseline drifts and incomplete resolution. In addition, computing integrators can also be programmed to perform other operations. If all the measurement parameters (slope sensitivity, etc.) are properly selected, these integrators provide the most accurate peak area measurements.

2 CALIBRATION METHODS

Since peak areas are proportional to the amount of substance injected, areas must be converted into concentration terms to be of use to the researcher. Several methods of quantification are used: external standard, internal standard, and standard addition.

2.1 External Standard Method

In the external standard method, a series of solutions of the reference compound are chromatographed and a plot of peak area or height versus concentration is constructed. If the system is operating properly, the calibration plot should result in a straight line (within a given concentration range) intercepting the origin. The slope of the linear part of the curve is the response factor, R:

$$R = \frac{\text{peak area (height)}}{\text{amount (concentration)}} \tag{139}$$

The response factor is used to calculate the amount of the component X in the unknown sample:

$$X = \frac{\text{area (or height)} \times \dfrac{1}{R} \times \text{total sample volume}}{\text{volume injected}} \qquad (136)$$

Better accuracy is achieved if calibration plots are constructed with samples of the sample matrix to which varying amounts of reference compound have been added. This eliminates possible interferences from the background components. Points on the calibration curve should be checked daily by chromatographing the standard solutions. If deviations are encountered, a new calibration plot should be constructed. The external calibration method is simple and affords good accuracy, provided the injection volumes are reproducible.

2.2   Internal Standard Method

The internal standard method involves the addition of a reference compound of known concentration to the unknown sample. Provided the internal standard is properly chosen, variations in the separation parameters are compensated for. It should be borne in mind, however, that since this procedure involves determinations of areas or heights for two peaks, the precision of the analysis is decreased by a factor of $1.4(\sqrt{2})$. However, the method is invaluable for analyses that require elaborate sample pretreatment, where considerable losses in recovery can result. For a compound to qualify as an internal standard, it must fulfill the following requirements:

1   It must be eluted in a vacant spot in the chromatogram.
2   It must be completely resolved from the neighboring peaks.
3   It must have a k' value similar to the k' of the peak of interest.
4   It must be chemically similar to the compound of interest.
5   It must be added at a concentration similar to the peak of interest.
6   It must be stable and available in high purity.
7   It must be absent in the sample.

If several compounds are to be determined in a sample, and if their k' values differ widely, more than one internal standard is necessary.

The calibration curve is constructed by chromatographing a series of solutions containing the compound of interest and the internal standard. Once the areas (or heights) of the unknown peak and internal standard have been computed, the ratios of their peak sizes are calculated and plotted versus the concentration of the compound of interest.

2.3 Standard Addition Method

The use of the standard addition method is advantageous in the analysis of complex mixtures when sample blanks cannot be obtained. This method involves chromatographing of the sample, followed by the analysis of another sample aliquot to which a known concentration of the compound of interest is added. The concentration of the compound of interest can be determined using the calibration factor, R, defined by the following equation:

$$R = \frac{h_x' - h_x}{W_x} \tag{141}$$

where h_x' is the height of the peak of interest, h_x is the height of the peak of the added standard in the original sample, and W_x is the weight of the compound added. The weight of the compound in the original sample is given by:

$$W_X = \frac{h_x}{R} \tag{142}$$

REFERENCES

1. R. A. Hartwick, A. M. Krstulović, and P. R. Brown, *J. Chromatogr.*, **186**, 659 (1979).

2. R. A. Hartwick, S. P. Assenza, and P. R. Brown, *J. Chromatogr.*, **186**, 647 (1979).

3. S. Chen, D. M. Rosie, and P. R. Brown, *J. Chromatogr. Sci.*, **15**, 218 (1977).

4. D. A. van Haverbeke and P. R. Brown, *J. Liquid Chromatogr.*, **1**, 507 (1978).

5. R. Keller, A. Oke, J. Mefford, and R. N. Adams, *Life Sci.*, **19**, 995 (1976).

6. T. M. MacNeil and P. R. Brown, Master's thesis, Univ. of R.I., 1979.

7. P. J. Naisch, R. E. Chambers, and M. Cooke, *Ann. Clin. Biochem.*, **16**(5), 254 (1979).

8. M. Israel, W. J. Pegg, P. M. Wilkinson, and M. B. Garnick, *J. Liquid Chromatogr.*, **1**, 795 (1978).

9. R. A. Hartwick and P. R. Brown, *J. Chromatogr.*, **126**, 679 (1976).

10. K. T. Koshy and A. L. van der Slik, *Anal. Biochem.*, **74**, 282 (1976).

11. N. E. Hoffman and J. C. Liao, *Anal. Chem.*, **49**, 2231 (1977).

12. F. K. Chow and E. Grushka, *J. Chromatogr.*, **185**, 361 (1979).

13. J. H. Knox and J. Jurand, *J. Chromatogr.*, **110**, 103 (1975).

14. F. S. Anderson and R. C. Murphy, *J. Chromatogr.*, **121**, 251 (1976).

15. A. M. Krstulović, R. A. Hartwick, and P. R. Brown, *Clin. Chem.*, **25**, 235 (1979).

16. P. Jandera and J. Churacek, *J. Chromatogr.*, **91**, 207 (1974).

17. P. J. Schoenmakers, H. A. H. Billiet, R. Tijssen, and L. DeGalan, *J. Chromatogr.*, **149**, 519 (1978).

18. H. Colin, N. Ward, and G. Guiochon, *J. Chromatogr.*, **149**, 169 (1978).

19. R. A. Hartwick, C. M. Grill, and P. R. Brown, *Anal. Chem.*, **51**, 34 (1979).

20. P. R. Brown, *J. Chromatogr.*, **52**, 257 (1970).

21. P. R. Brown, *High Pressure Liquid Chromatography, Biochemical and Biomedical Applications*, Academic, New York, 1973.

22. P. R. Brown, A. M. Krstulović, and R. A. Hartwick, *Human Her.*, **27**, 167 (1977).

23. R. Yost, J. Stoveken, and W. MacLean, *J. Chromatogr.*, **134**, 73 (1977).

24. A. M. Krstulović, P. R. Brown, and D. M. Rosie, *Clin. Chem.*, **23**, 1984 (1977).

25. A. M. Krstulović, R. A. Hartwick, P. R. Brown, and K. Lohse, *J. Chromatogr.*, **158**, 365 (1978).

26. A. M. Krstulović, P. R. Brown, D. M. Rosie, and P. B. Champlin, *Clin. Chem.*, **23**, 1984 (1977).

27. T. Jupille, *J. Chromatogr. Sci.*, **17**, 160 (1979), and references contained therein.

28. P. T. Kissinger, G. Bratin, G. C. Davis, and L. A. Pachla, *J. Chromatogr. Sci.*, **17**, 137 (1979), and references contained therein.

29. J. F. Lawrence, *J. Chromatogr. Sci.*, **17**, 147 (1979), and references contained therein.

30. S. Ahuja, *J. Chromatogr. Sci.*, **17**, 168 (1979).

31. J. F. Lawrence and R. W. Frei, *Chemical Derivatization in Liquid Chromatography*, Elsevier, Amsterdam, 1976.

32. K. Blau and G. J. King, *Handbook of Derivatives for Chromatography*, Heyden, London, 1977.

33. R. S. Deelder, M. G. F. Kroll, A. J. B. Beeren, and J. H. M. van den Berg, *J. Chromatogr.*, **149**, 669 (1978).

34. R. S. Deelder, M. G. F. Kroll, and J. H. M. van den Berg, *J. Chromatogr.*, **125**, 307 (1976).

35. L. J. Skeggs, *Am. J. Clin. Pathol.*, **28**, 311 (1957).

36. L. R. Snyder and H. J. Adler, *Anal. Chem.*, **48**, 1017, 1022 (1976).

37. L. R. Snyder, *J. Chromatogr.*, **125**, 287 (1976).

38. R. W. Deelder and P. J. H. Hendricks, *J. Chromatogr.*, **83**, 343 (1973).

39. R. W. Frei and A. H. M. T. Scholten, *J. Chromatogr. Sci.*, **17**, 152 (1979).

40. N. Tanaka and E. R. Thornton, *J. Am. Chem. Soc.*, **99**, 7300 (1977).

41. I. Molnár and C. Horváth, *J. Chromatogr.*, **142**, 623 (1977).

42. N. Haddon et al., *Basic Liquid Chromatography*, Varian Aerograph, Walnut Creek, CA, 1971.

43. P. R. Brown, A. M. Krstulović, and R. A. Hartwick, *J. Clin. Chem. Biochem.*, **14**, 282 (1976).

44. R. A. Hartwick and P. R. Brown, *J. Chromatogr. Biomed. Appl.*, **143**, 383 (1977).

45. P. R. Brown, R. A. Hartwick, and A. M. Krstulović, in *Biological/Biomedical Applications of Liquid Chromatography*, Vol. II, G. L. Hawk, Ed., Dekker, New York, 1979.

46. A. M. Krstulović, P. R. Brown, and D. M. Rosie, *Anal. Chem.*, **49**, 2237 (1977).

47. A. M. Krstulović and A. M. Powell, *J. Chromatogr.*, **171**, 345 (1979).

48. A. M. Krstulović, M. Zakaria, K. Lohse, and L. B. Dziedzic, *J. Chromatogr.*, **186**, 733 (1979).

48a. A. M. Krstulović, L. B. Dziedzic, S. W. Dziedzic, and S. E. Gitlow, *J. Chromatogr. Biomed. Appl.*, **223**, 305, (1981).

48b. A. M. Krstulović and P. R. Brown, *Am. Lab.* **17**, May 1981.

49. R. A. Hartwick, D. vanHaverbeke, M. McKeag, and P. R. Brown, *J. Liq. Chromatogr.*, **2**, 725 (1979).

50. L. R. Snyder and J. J. Kirkland, *Introduction to Modern Liquid Chromatography*, 2nd ed., Wiley-Interscience, New York, 1979.

51. R. P. Brown and E. Grushka, *Anal. Chem.*, **52**, 1210 (1980).

52. M. Uziel, C. K. Koh, and W. E. Cohn, *Anal. Biochem.*, **25**, 77 (1968).

53. C. G. Horváth, B. A. Preiss, and S. R. Lipsky, *Anal. Chem.*, **39**, 1422 (1967).

XI Selected Biochemical/ Biomedical Applications

The development of RPLC has revolutionized the analyses of biological molecules. The majority of biochemically important compounds are thermally labile and nonvolatile; thus they cannot be separated by gas GLC without derivatization. However, with RPLC, both closely related and highly diverse compounds can be analyzed simultaneously with minimal sample preparation. In addition to nonpolar and polar molecules, ionized and ionizable molecules can also be analyzed using ion-association and ion-suppression techniques.

RPLC has been used in biochemical and biomedical research since the early 1970s, but it has only recently made its way into the clinical laboratory. Its use has made possible rapid analyses of drugs and their metabolites not only for determining compliance and therapeutic drug monitoring, but also for the determination of drugs of abuse in forensic and toxicology laboratories. It has been found that serum levels of a therapeutic drug, or one of its metabolites, may correlate more closely to clinical efficacy than the prescribed dosages because of individual variation in total rate of drug metabolism. Many factors, such as diet, general health, and medications taken, affect the serum levels of certain drugs. Thus it is especially important to be able to monitor, over long periods of time, the serum levels of drugs used to treat patients with both acute and chronic diseases, as well as their urinary metabolites.

Qualitative and quantitative analyses of naturally occurring constituents of physiological fluids, solid wastes, and body tissues can be carried out rapidly and sensitively. Separations are highly reproducible, and with most detectors, the samples are not destroyed. RPLC can also be used as a diagnostic tool, since many compounds can be determined in one assay; thus metabolic profiles of naturally occurring constituents in physiological fluids can be obtained and used as biochemical markers of disease states, for monitoring the progress of various physical and mental disorders, and for the determination of the effectiveness of therapy and/ or the onset of relapse.

218

Therefore, RPLC can be used in pharmacological, toxicological, forensic, and clinical laboratories. However, it is not limited to clinical applications. It can be used to great advantage in the basic sciences, for example, in basic biochemical studies involving metabolic pathways, cell growth, reproduction, and disease and regenerative processes, and in tissue culture work, bacteriology, virology, immunology, medicinal chemistry, pathology, and genetics. In addition, its use is tremendously useful in kinetic studies and in determining thermodynamic parameters. RPLC is also used for monitoring synthetic organic reactions and in the analysis and isolation of reaction intermediates and products.

Because of the "explosion" of applications of HPLC in biochemistry and clinical chemistry in the past few years, we will only highlight some of the recent advances in this field. Since many RPLC assays of biologically active compounds are reported in the literature every month, our objective will be to "whet the appetite" of the readers in the hope that they will investigate in depth their own area of interest.

1 AMINO ACIDS, PEPTIDES, AND PROTEINS

Research involving amino acids, peptides, and proteins, which occur in trace amounts in biological systems, has increased our understanding of many complex biochemical and endocrine pathways. For example, recently discovered brain opiatelike peptides, encephalins, and endorphins are involved in fundamental brain functions and appear to act as neurotransmitters of specific pathways that process information relating to pain and emotional behavior. Because of the great interest in the naturally occurring low-molecular-weight peptides and proteins, there is a tremendous need for rapid separations and sensitive detection methods to monitor these compounds at their endogenous levels.

Classically, amino acids in proteins have been separated by automatic amino acid analyzers (1) which involve postcolumn derivatization. Both ion exchange (2–4) and RP packings (5–15) have been used in the analysis of these compounds; however the RP mode now appears to be the method of choice. Although C_{18} columns are widely used, Grushka et al. (13–13b) reported on the separation of amino acids as well as peptides on specially prepared bonded peptide phases.

The development of RPLC as an analytical tool opened up new possibilities in the analysis of complex mixtures of di- and polypeptides. Early RP analyses had many limitations: resolution was poor, peaks were broad, and retention times were long. Furthermore, with the polypeptides and proteins, recoveries were low due to the irreversible binding to the column through hydrophobic interactions (3). However, ion-association chromatography has improved the analysis of peptides and proteins, and the use of hydrophilic and/or hydrophobic pairing agents has been found advantageous for the analysis of complex mixtures (15). To improve the separation and detectability of amino acids, peptides, and proteins, derivatization has traditionally been used with the chromatographic separations. Usually, amino acids are analyzed as their dansyl (16) or phenylthiohydantoin (PTH) derivatives (17). An example of the analysis of PTH amino acids is shown in Fig. 83. To increase the sensitivity and selectivity of the analysis of amino acids, Hill et al. (18) developed a method in which fluorescent derivatives of the acids were analyzed. They separated the o-phthaldehyde ethandiol derivatives of 20 common amino acids in human serum using a fluorescence detector. Molnár and Horváth (14) used the ion-association technique to separate some amino acids (Fig. 84). In addition, they separated peptides using RPLC and detected them without derivatization (Fig. 85). Rivier (11) showed that the use of trialkyl ammonium phosphate buffers in RPLC gives high recoveries and resolution for peptides and proteins. This system can be used for peptide mapping and the determination of optical and chemical purity of oligopeptides and polypeptides.

RPLC has also been successful (Fig. 86) in resolving peptide hormone diastereoisomers (12). Grushka and co-workers (13) have found that complexing agents, such as Cu^{2+} with aspartam, are very effective aids in separating diastereoisomers of amino acids. This technique in separating diastereoisomeric mixtures of peptides, peptide derivatives, and analogs, both analytically and preparatively, has great potential in biochemical and synthetic work.

2 BIOGENIC AMINES AND RELATED COMPOUNDS

Biogenic amines are ubiquitous and are believed to play a key role in cell biology. For example, catecholamines (CA), a class of compounds with a 3,4-dihydroxyphenyl ring and a side chain of ethylamine or ethanolamine, act as neurotransmitters in the adrenergic system. The discovery that CA are found within specific neural pathways has stimulated research on their physiological role in normal behavior as well as in

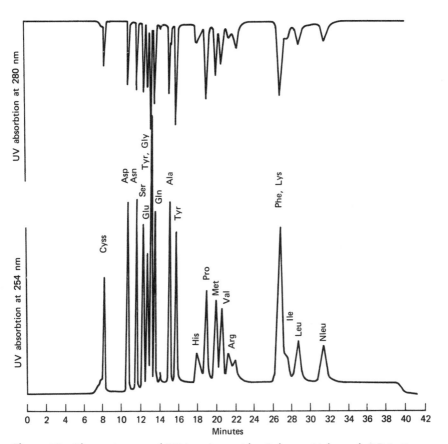

Figure 83 Chromatogram of PTH amino acids. Column, Lichrosorb RP-8, 7 μm (Knauer); mobile phase: Eluent A, 1.05 M acetate buffer (pH 4.1); Eluent B, CH₃OH/ A, (90:10) (%); gradient from A to B in 10 min; flow rate, 1.5 ml/min; temperature, 28°C. Reproduced with permission of Dr. H. Lehmann, MPI Molecular-biologie, Berlin, and Dr. H. Knauer.

Parkinson's disease (19,20), shock and stress (21), the process of learning (22), sleep (23), and regulation of body temperature (24). In addition, since the discovery of the metabolic pathways of CA by Armstrong et al. (24a), the determination of CA, their precursors, and major metabolites has become considerably important in the understanding of essential hypertension (24b), muscular distrophy (24c), and neural crest tumors, such as pheochromocytoma, neuroblastoma, and ganglioneuroma (24d,48). The clinical and histopathologic diagnosis of neural crest tumors is pla-

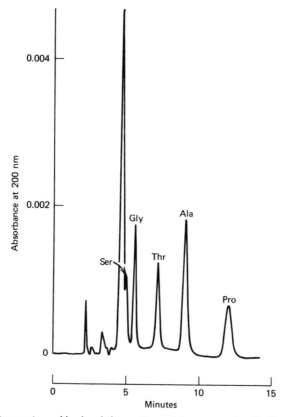

Figure 84   Separation of hydrophilic amino acids on octadecyl-silica column with decyl sulfate in the neat aqueous eluent. Column, Lichrosorb RP-18, 5 μm; eluent, 0.1 M phosphate buffer (pH 2.1 containing 3×10^{-3} M decyl sulfate); temperature, 70°C; flow rate, 2 ml/min; ΔP, 150 atm. Reproduced from reference 14 with permission.

gued with difficulties due to their small size, intra-abdominal position, and symptoms similar to those of essential hypertension. However, quantitative analysis of urinary CA and their catabolites has given the clinician a unique tool to diagnose these life-threatening tumors reliably and rapidly and to differentiate between patients with these tumors and the vast majority of those whose hypertension is of different etiology.

The deactivation of the three principal CA, noreprinephrine (NE), epinephrine (E), and dopamine (D), proceeds via 3-O-methylation by the

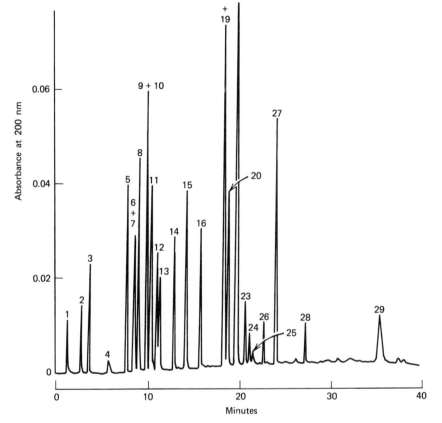

Figure 85 Chromatogram of nonpolar amino acids and small peptides. Column, Lichrosorb RP-18, 5 μm; gradient elution from 0.5 M HClO₄ (pH 0.2), with acetonitrile as the gradient former; temperature, 70°C; flow rate: 2 ml/min; initial ΔP, 150 atm. Sample: 10 μl containing ca. 1 μg of each component. 1, Glu; 2, Val-Ala; 3, Tvr; 4, Leu; 5, Phe-Gly-Gly; 6, Phe; 7, Phe-Gly; 8, Trp-Glu; 9, Trp-Gly; 10, Gly-Phe; 11, Trp-Ala; 12, Trp; 13, Gly-Trp; 14, Ala-Phe; 15, Trp-Tyr; 16, Val-Ala-Ala-Phe; 17, Phe-Phe; 18, Phe-Gly-Gly-Phe; 19, Phe-Gly-Phe-Gly; 20, Trp-Leu; 21, Trp-Phe; 22, Trp-Trp; 23, Trp-Met-Asp-Phe-NH₂; 24, Unknown; 25, Unknown; 26, Lys-Phe-Ile-Gly-Leu-Met; 27, Phe-Phe-Phe; 28, Phe-Phe-Phe-Phe; 29, Phe-Phe-Phe-Phe-Phe. Reproduced from reference 14 with permission.

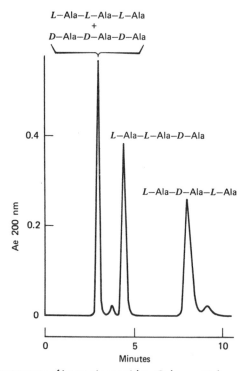

Figure 86   Chromatogram of isomeric peptides. Column, Lichrosorb RP-18, 10 μm, av. particle size (250 × 46 mm i.d.); mobile phase, 0.1 M phosphate buffer (pH 2.1); flow rate: 2.0 ml/min; temperature, 25°C. Reproduced from reference 12 with permission. Copyright (1978) American Chemical Society.

enzyme catechol-O-methyltransferase (COMT) and monoamine oxidase (MAO). Dopamine catabolism gives rise to homovanillic acid (HVA), whereas norepinephrine and eprinephrine are degraded centrally and peripherally to 3-methoxy-4-hydroxyphenylethylene glycol (MHPG) and vanillylmandelic acid (VMA), respectively.

The di- and polyamines such as 1,3-propanediamine, putresceine, cadaverine, spermidine and spermine, and their acetylated derivatives are of special interest, since it has been found that elevated levels of these compounds occur in serum (25), urine (26), and cerebrospinal fluid (27) of patients with metastatic cancer. Since there appears to be a direct correlation between the polyamine concentration and the clinical status of the patients, it has been postulated that the polyamine levels in physiological fluids could be used as biological markers to diagnose and monitor patients with cancer. Furthermore, polyamines have been found in prokaryotic and eukaryotic cells as well as in virus particles. Because of the exceedingly low levels at which biogenic amines and their metabolites occur in biological fluids, highly sensitive analytical methods with selective detection systems are required in physiological and pharmacological research on these compounds.

Most biogenic amines are highly sensitive to light and oxygen, decompose readily on exposure to air, and undergo spontaneous oxidation at alkaline pH values. However, in contrast to open column chromatography, oxidation and exposure to light are minimized in the RPLC analysis of these compounds. This, in addition to its speed and sensitivity, makes RPLC the method of choice for these assays.

For the analysis of amines in tissue, samples are usually homogenized in perchloric acid. The amines are then concentrated on a sorbent, such as alumina or boric acid gel, after the adjustment of pH. Adsorbed compounds are then eluted with a suitable solvent.

The determination of CA in the central nervous system usually requires the use of invasive techniques, whereas the assessment of trace amounts of serum and urinary CA necessitates elaborate sample pretreatment techniques and highly sensitive and time-consuming analytical methods. GLC with electron capture or flame ionization detection alone or in tandem operation with MS offers great sensitivity and specificity, but requires sample derivatization to increase the volatility of these thermally labile

Different modes of HPLC have been used for the analysis of these compounds: ion exchange (28,29), adsorption (30–32), RP ion association (33–35), and more recently, direct RPLC analysis (36–39). Since the first two techniques did not provide the separation of both the acidic and the basic sample constituents simultaneously, the reversed phase, both with and without ion-association reagents, is mainly used. The main problem in these assays has been the need for adequately sensitive detection devices, since the commonly used UV-absorbance detectors for HPLC do not provide the sensitivity needed in the analysis of endogenous amines. However, higher sensitivity can be achieved by fluorometric detection, and nanogram amounts of CA can easily be detected.

Although measurements of native fluorescence are sufficiently sensitive

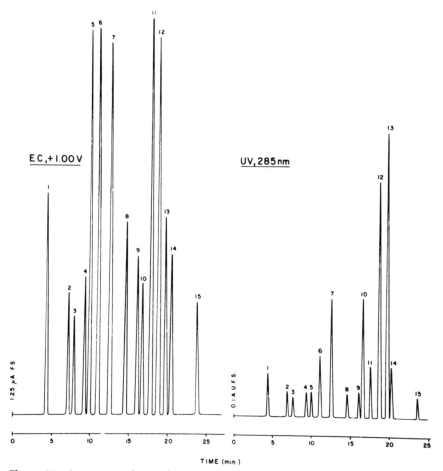

Figure 87 Separation of a synthetic mixture containing approximately 2 nm each of the following: (1) 3,4-dihydroxymandelic acid; (2) dopa; (3) methanephrine; (4) tyrosine; (5) vanillylmandelic acid; (6) 3-methoxy-4-hydroxyphenylethyleneglycol; (7) 5-hydroxytryptophan; (8) 3,4-dihydroxyphenylacetic acid; (9) anthranilic acid; (10) tryptophan; (11) 5-hydroxyindole-3-acetic acid; (12) vanillic acid; (13) 3,4-dihydroxycinnamic acid; (14) 3-indoleacetamide; (15) 3-indole lactic acid. Chromatographic conditions: column, μBondapak/C$_{18}$; (low strength) eluent, 0.1 F KH$_2$PO$_4$ (pH 2.50); (high strength) eluent, acetonitrile-water (3:2); gradient, linear from 0 to 60% of the high strength eluent in 45 min; flow rate, 1.4 ml/min; temperature, ambient; detection, electrochemical at +1.00 V vs. Ag/AgCl and UV absorption at 285 nm. Reproduced from reference 49 with permission.

for the analysis of tissue CA, derivatization is mandatory if CA are to be analyzed in biological fluids. Improved detectability can be achieved by derivatization with fluorescamine (40,41), dansyl chloride (42–44), o-pthaldehyde (45), catechol oxidation and alkaline rearrangement (46), or reaction with 2,5-dimethyltetrahydrofuran and p-dimethylamino benzaldehyde (47).

However, the recent development of the electrochemical detectors has eliminated the need for derivatization. The amperometric detection of CA exploits the two-electron oxidation of amines to the corresponding o-quinones, and the resulting current is proportional to the concentration of the electroactive species. Figure 87 illustrates the RP separation of a synthetic mixture of CA metabolites and precursors, detected both amperometrically and spectrophometrically. High sensitivity afforded by the amperometric detection makes this technique ideally suited for the determination of the excretion patterns of CA degradation products. The use of this analytical procedure is illustrated in the analysis of VMA and free MHPG in a urine sample from a healthy subject (Fig. 88) and a patient with clinically diagnosed pheochromocytoma (Fig. 89). Because of the relative nonspecificity of amperometric detection, phenolic acids are usually separated from the basic and neutral compounds by means of an extraction with ethyl acetate. This simplifies the analysis of CA metabolites and makes possible the detection at higher oxidation potentials without a concomitant loss of specificity.

3 ENZYMES

Determinations of enzyme activities are being used in the diagnosis of several disease states. The analytical procedures are usually based on spectrophotometric analyses, which often lack specificity and require large sample volumes.

The use of RPLC has great potential for the measurement of enzyme activities in body fluids and tissues. For these assays, substrate and/or product concentrations can specifically be monitored by RPLC. The procedure includes incubation of the sample with a suitable substrate and termination of the reaction by heat, addition of acid, or a specific inhibitor. The change in substrate and/or product concentration is measured by RPLC and the enzymatic activity is calculated. RPLC offers several advantages for enzyme assays: the specificity afforded by the separation

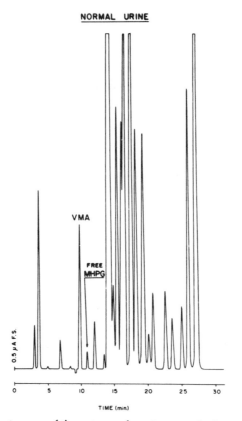

Figure 88 Chromatogram of the extract of a urine sample from a healthy subject. Chromatographic conditions are the same as in Fig. 87. Detection, electrochemical at +1.00 V vs. Ag/AgCl; volume of extract injected, 5 μl. Reproduced from reference 49 with permission.

process, the high sensitivity of RPLC, and the low sample volumes required.

Pennington (49a) was one of the first to use HPLC for the measurement of enzymatic activities. Using the ion-exchange mode, he determined 3',5'-cyclic adenosine monophosphate phosphodiesterase activity by monitoring the change in concentration of cAMP in a sample. Uberti et al. (50) also used the ion-exchange mode to determine the activity of adenosine deaminase, as did Nelson (51), to measure the activities of purine nucleoside phosphorylase, CDP-choline synthetase, and β2'-deoxythioguanosine kinase.

Hartwick et al. (52) used RPLC to determine adenosine deaminase activity in erythrocytes by monitoring adenosine concentrations. Compared with other techniques, this assay minimizes interferences caused by endogenous enzymes. Furthermore, it enables simultaneous monitoring of products of several competing enzymatic reactions. Figure 90 shows the determination of the adenosine deaminase activity in human erythrocytes using RPLC. Activities measured by this method were found to correlate well with those measured by other standard methods. This technique is also useful in determining optimal conditions for the incubation procedure as is shown in Fig. 91.

Krstulović et al. (53) have devised an RPLC assay for the measurement of acid and alkaline phosphatase activities in serum. In this assay, Ni^{2+} ions were used to inhibit 5'-nucleotidase selectively, which would otherwise interfere with the measurement of phosphatase activity.

Using the principles employed for these enzymatic assays, other RPLC enzymatic assays are being developed. With the increasing use of RPLC as a general purpose instrument in the clinical and biomedical laboratories,

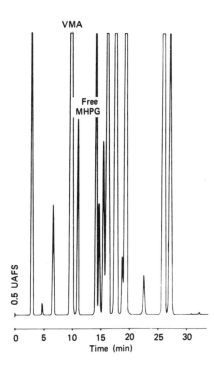

Figure 89 Chromatogram of the urine extract of a sample from a patient with clinically diagnosed pheochromocytoma. Chromatographic conditions are the same as in Fig. 87, detection, electrochemical at +1.00 V vs. Ag/AgCl; volume of extract injected, 5 μl. Reproduced from reference 49 with permission.

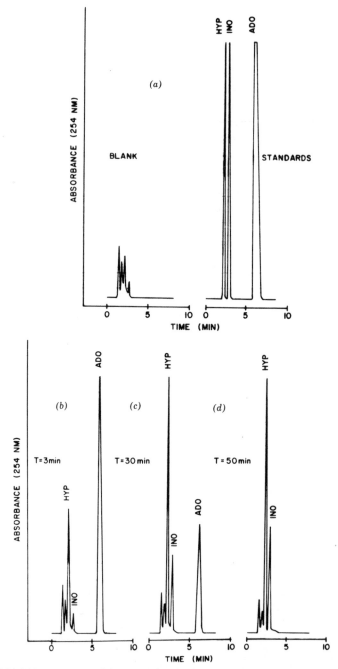

Figure 90 Measurement of the activity of adenosine deaminase in human erythrocytes using HPLC. In (a) a blank erythrocyte lysate is shown along with three standards: hypoxanthine (Hyp), inosine (Ino), and adenosine (Ado). In (b), (c) and (d) the decrease in the substrate area (Ado) is shown as a function of incubation time with the erythrocytes. Column, μBondapak/C_{18}; mobile phase, 86/14 (v/v) 0.01 M KH_2PO_4/CH_3OH; (86:14 v/v) flow rate, 2.0 ml/min; detection wavelength, 254 nm; temperature, ambient. Reproduced from reference 52 with permission of Preston Publications Inc.

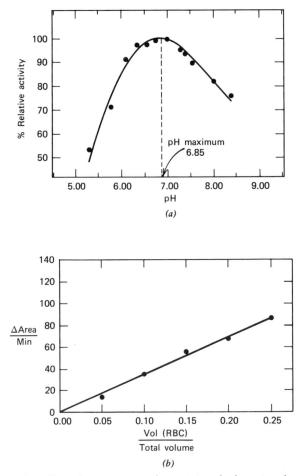

Figure 91 (a) The affect of assay pH on the activity of adenosine deaminase. The ionic strength of the incubation medium was held approximately constant by making the incubate 0.05 *F* in KCl and 0.025 *F* in phosphate. The pH was adjusted by the addition of KOH or HCl. (b) Adenosine deaminase activity as a function of the volume of erythrocytes, 200 μl of water and 1.00 ml of stock Ado) was chosen for the final protocol conditions. Reproduced from reference 52 with permission of Preston Publications Inc.

it is predicted that many more rapid and selective enzymatic assays will soon become available.

4 LIPIDS

Lipids are a heterogeneous class of compounds that are characterized by their solubility in organic solvents and insolubility in water. They are divided into four principle classes: fats and oils, phosphoglycerides, sphingolipids, and steroids. Fats and oils include the common fatty acids, which are mainly straight-chain saturated or unsaturated acids. The phosphoglycerides are defined as any lipid derived from glycero-phosphoric acids that contain two long-chain fatty acyl chains. They include phosphatidyl cholines (called lecithins), the ethanolamines, and serines. The sphingolipids, which contain a sphingosine base or its derivative, and the steroids discussed in Section 6, all have a perhydrocyclopentanophenanthrene ring system.

Although lipids are very important biochemically, there are few rapid and efficient methods to separate and identify the classes of lipids as well as the individual molecular species within each class. Classically, a form of liquid-solid chromatography, argentation, was mainly used for separations of fatty acids, derivatives of fatty acids, and alcohols and triglycerides. Recently, RPLC procedures have been developed, and some of them are outlined in Table 1 (54–62). Analytical methods are now being developed to analyze separate species of phosphoglycerides.

In 1976, McCluer and Jungalwola (63) were able to separate some synthetic phosphatidic acids by making the methyl esters and using a μBondapak C_{18} column with a buffered methanol eluent. They used a variable wavelength detector and found that the separation was based on chain length as well as on the degree of unsaturation. Using two ODS columns in series and an acetonitrile eluent, Hsieh et al. (55) also developed a method for separating dimethyl esters of phosphatidic acids and detected them using a RI detector. They were able to obtain baseline separation of eight of the nine major phosphatidic acids that occur in egg yolk and soybean (Fig. 92). A positive linear relationship was obtained when the total carbon number of the fatty acid chains was plotted against log retention time and a negative slope when the number of double bonds was plotted against log retention time (Fig. 93). Porter et al. (64) also used two RP columns, but they used them independently to separate 10

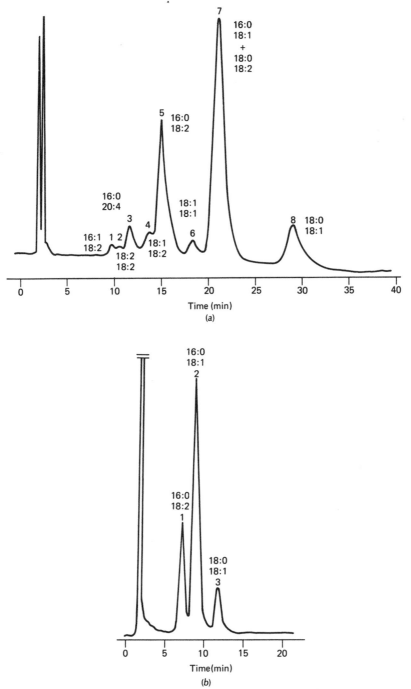

Figure 92 Separation of phosphatidic acid dimethyl esters (DADE) derived from egg yolk phosphatidycholine. (a) Columns, two Partisil-10 ODS in series; mobile phase, acetonitrile; flow rate, 3 ml/min at 900 psig; detection, RI × 16. (b) Column, Partisil-10 ODS-2; mobile phase, methanol; flow rate, 2 ml/min at 1250 psig; detection, FID × 16. Reproduced from reference 58 with permission.

233

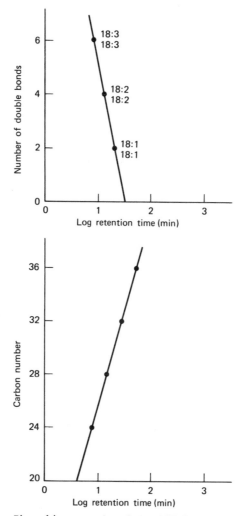

Figure 93 Upper: Plot of log retention times (HPLC) versus carbon numbers of phosphatidic acid dimethyl esters (PADE) having different degrees of unsaturation. Lower: Plot of log retention times (HPLC) versus carbon numbers of saturated synthetic phosphatidic acid dimethyl esters (PADE). Reproduced from reference 58 with permission.

different synthetic phosphatidylcholines. Although C_{18} columns have been used most commonly for these separations, C_8 packings generally give as good or better results in terms of column efficiency, resolution, and analysis time. The use of ion association has also been investigated for the separation of molecular species of intact phosphoglycerides and is applicable especially to groups containing zwitterions.

The major problem in the analysis of lipids by RPLC is the method of detection, since most of these compounds lack strongly absorbing UV chromophores. The two detection methods commonly used are the flame ionization detector, as shown in Fig. 92b, and the RI detector, as shown in Fig. 92a. However, UV detection can be used if suitable derivatives are formed. Naphthacyl, phenacyl, methoxyanilides, bromophenacyl, and methoxyphenacyl derivatives of fatty acids or phosphoglycerides have all been prepared successfully and detected spectrophotometrically. Derivatives of methylmethoxycoumarin were also used by Lam and Grushka (66) to separate fatty acids (Fig. 94). Promise for the future for better detection and identification of lipids lies in the interfacing of MS with RPLC (67).

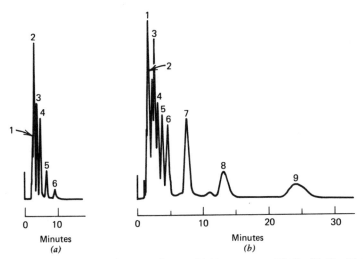

Figure 94 (a) Separation of C_1—C_6 fatty acid Mme esters. (1) C_1; (2) C_2; (3) C_3; (4) C_4; (5) C_5; (6) C_6. Detector sensitivity, 50; mobile phase, 15% water in methanol; flow rate, 1.96 ml/min. (b) Separation of C_6—C_{16} fatty acid Mme esters. (1) Decomposed derivatization reagent; (2) C_6; (3) C_7; (4) C_8; (5) C_9; (6) C_{10}; (7) C_{12}; (8) C_{14}; (9) C_{16}. Detector sensitivity, 50; mobile phase, 15% water in methanol; flow rate, 1.96 ml/min. Reproduced from reference 66 with permission.

5   NUCLEOTIDES, NUCLEOSIDES, AND THEIR BASES

Because of the significance of nucleic acids in biochemistry, biology, medicine, and other fields of research, there has been a great demand for the development of new or improved methods for the analysis of nucleic acid components. Early work on HPLC was spurred by the need for rapid methods for determining base pair composition of hydrolysates of DNA and RNA (67,68). In addition, the concentrations of free nucleotides, nucleosides, and bases in biological samples were found to be of great interest for investigating metabolic states, normal metabolism of nucleic acids, and abnormalities caused by diseases.

With the development of HPLC, accurate and rapid high-resolution analyses are now possible for the determination of major and modified nucleosides in nucleic acids (69) and small DNA and RNA oligo nucleotides (70). Moreover, HPLC has made possible the determination of the naturally occurring nucleotides, nucleosides, and their bases in very small samples of tissue, urine, blood, and other physiological fluids (72–76,78–84).

In the early days of HPLC, nucleotides that contain one or more phosphate groups were separated by ion exchange. However, using ion-association agents, nucleotides can now be separated by RPLC (76). Since the retention of nucleotides is governed by the combination of solvophobic ion concentration, ionic strength, and pH, it is possible to achieve good separations by changing one or more of the separation parameters. Most often, a large counter ion such as the tetrabutyl-ammonium ion is added to the mobile phase. The resulting complex, which consists of an ion pair, is effectively nonionic and can thus be chromatographed by RP chromatography. Hoffman and Liao (76) reported on the separation of 12 nucleoside-5'-phosphates and cyclic AMP by the ion-association technique (Fig. 56). The order of elution can be rationalized in terms of charge and stationary RP interaction.

The RP mode of HPLC is particularly suited for the separation of nucleosides and bases (73,74). Using a microparticulate C_{18} column and a water-methanol gradient, rapid and sensitive analyses of nucleosides and bases can be achieved with good resolution (Fig. 95). In addition, the nucleotides present in biological extracts do not interfere with the analysis, since most of them are eluted before the nucleosides and bases. Serum profiles of these compounds, along with other UV-absorbing, low-molecular-weight compounds, in a patient with renal failure and a patient

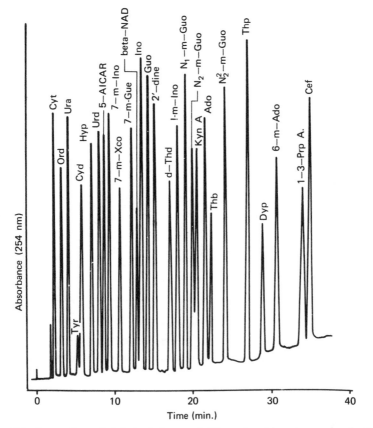

Figure 95 Separation of 0.1 to 0.5 nm of 28 nucleosides, bases, nucleotides, aromatic amino acids, and metabolites. Injection volume 40 μl of a solution 1 × 10^{-5} mol/l in each standard. Column: chemically bonded reversed phase (C$_{18}$) on 10μm totally porous silica support. Low strength eluent, 0.02 mol/l KH$_2$PO$_4$ (pH 5.6); high strength eluent, 60% methanol; gradient, slope 0.69%/min (0 to 60% methanol in 87 min), linear; temperature, ambient; flow rate, 1.5 ml/min. Reproduced from reference 73 with permission.

with breast cancer are shown in Figs. 96 and 97, respectively. Since significant differences were found in the chromatograms of some patients with various types of neoplastic diseases, the possibility of using RPLC serum profiles to detect disease states and monitor the progress of the disease and/or chemotherapy is now being investigated (73,74).

RPLC can also be used to investigate the effects of diet, smoking, air pollutants, and so forth on the serum profiles of UV-absorbing, low-

RENAL FAILURE

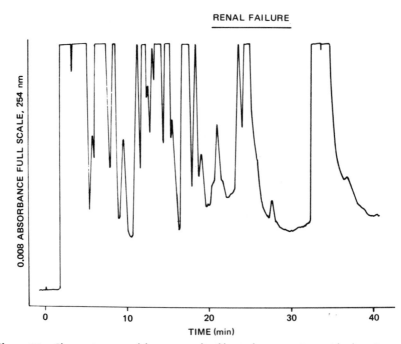

Figure 96 Chromatogram of the serum ultrafiltrate from a patient with chronic renal failure and chronic lymphocytic leukemia (CLL). Injection volume, 80 μl. Chromatographic conditions are the same as those given in Fig. 95. Reproduced from reference 73 with permission.

molecular-weight compounds. In a recent study with dogs it was found that smoking significantly affects the concentrations of compounds in the purine metabolic pathway and two essential amino acids, tryptophan and tyrosine (83,84) (Fig. 98).

By varying the operating conditions, analyses can be optimized for different groups of compounds or for a specific compound using the same column. For example, for studies of purine metabolism, the separation of purines and their nucleosides was optimized in the presence of pyrimidine compounds and nucleotides by changing the mobile phase (80). In other medical studies, such as in investigations of heart disease or adenosine deaminase deficiency, a rapid analysis of adenosine was required. This was accomplished isocratically using the same RPLC column and increasing the amount of methanol in the mobile phase (Fig. 99) (81).

It is also possible to devise selective analyses for a particular methylated nucleoside or a specific group of methylated nucleosides, such as 1-methylinosine and N^2-methylguanosine, in the presence of the other normally occurring compounds (84a). Using a highly selective affinity gel-isolation technique, Gehrke and co-workers (71–72) developed a selective analysis for methylated ribonucleosides, which were found to be elevated in urine from patients with neoplastic diseases. They used this method for determining the major and modified nucleosides in hydrolysates of t-RNA, after a boronate gel-filtration procedure (Fig. 100).

In addition, the adenine nucleotides can be analyzed along with aden-

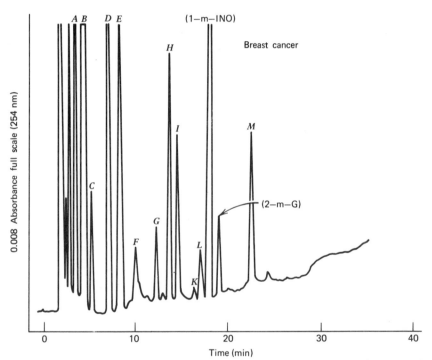

Figure 97 Serum profiles of a postoperative breast cancer patient with metastasis to the bone. Injection volume, 80 μl. Peaks identified with parentheses are tentatively based on retention times alone. Chromatographic conditions are the same as in Fig. 95. Peak identity: A, creatinine; B, uric acid; C, typosine; D, hypoxanthine; E, uridine; F, unknown; G, unknown; H, inosine; I, guanosine; K, hippuric acid; L, tryptophan; 1-methylinosine; 2-methylguanosine; M, thiobromine. Reproduced from reference 73 with permission.

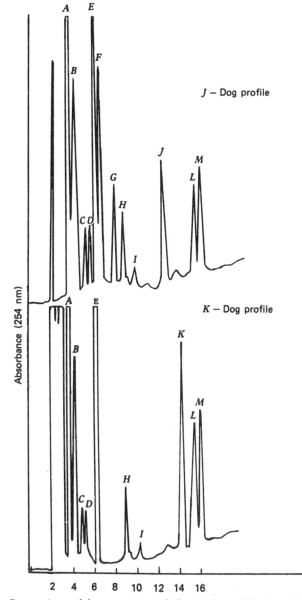

Figure 98 Comparison of the serum (J) and plasma (K) profiles found in two groups of dogs. Injection volume, 40 μl. Detector sensitivity, 0.02 a.u.f.s. Chromatographic conditions, μBondapak/C$_{18}$ (3.9 × 300 mm); low strength eluent, 0.02 M KH$_2$PO$_4$, (pH 5.7); high strength eluent, CH$_3$OH-H$_2$O (3:2, v/v); gradient, linear from 0 to 40% of the high strength eluent in 35 min; flow rate, 1.5 ml/min, temperature ambient. Reproduced from reference 84 with permission.

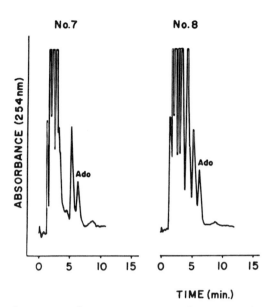

Figure 99 Sample Nos. 7 and 8 are serum extracts from two patients suffering from adenosine deaminase deficiency. 45 and 75 pico-moles are contained under the adenosine peak of sample Nos 7 and 8, respectively. Chromatographic conditions: column, Waters μBondapak/C$_{18}$; temperature, ambient; detector sensitivity, 0.02 a.u.f.s.; integrator setting 2; eluent, 90% 0.007 F KH$_2$PO$_4$ (pH 5.8) and 10% anhydrous methanol; flow rate, 2.0 ml/min. Reproduced from reference 81 with permission.

osine and adenine using RPLC(84b). Thus the concentrations of all the adenine compounds in the metabolic pathway can be determined simultaneously (Fig. 101).

6 STEROIDS

The steroids are a complex group of compounds that contain a perhydrocyclopentanophenanthrene ring system. The naturally occurring steroids are of critical importance to the body chemistry. Steroids are divided into four broad classes, according to their chemical properties and their biological effects: the corticosteroids, androgens, estrogens, and progestins.

Since the earliest separation of steroid hormones by HPLC in 1970 by Siggia and Dishman (88), extensive work has been done on the devel-

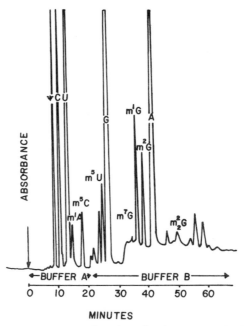

Figure 100   RP HPLC separation of nucleosides from t-RNA hydrolysate after boronate gel isolation; sample, 3 μg of yeast t-RNA. Chromatographic conditions: column μBondapak/C_{18} (600 × 4 mm); buffers, K, 2.5% (v/v) methanol in 0.01 M $NH_4H_2PO_4$ (pH 5.10); B, 10% (v/v) methanol in 0.01 M $NH_4H_2PO_4$ (pH 5.10); flow rate, 1.0 ml/min; detector, 254 mm: temperature, 37.6°C. Peak identity: ψ, pseudourine; C, cytidine; U, uridine; mA, 1-methyladenosine; m^5C, 5-methylcytidine; m^7G, 7-methylguanosine; I, inosine; m^5U, 5-methyluridine; G, Guanosine; m^1G, 1-methylguanosine; m^2G, 2-methylguanosine; A, adenosine; m^2_2G, N^2,N^2-dimethylguanosine. Reproduced from reference 72 with permission.

opment of HPLC assays for various classes of steroids (for review articles see references 85–87). Adsorption was first used for steroid separations (89–93); however, severe peak tailing was often encountered. Separations were also performed on stationary phases such as CTFE tetrafluoroethylene polymers and Amberlite LA-1 (94). With the recent trend toward RPLC, packings such as C_{18} (95–101), C_8 (102), and phenyl (103) are now being used. An example of an isocratic RP separation of some urinary estrogens, which are valuable indicators of health and organ status in females, is shown in Fig. 102. In the prevention and treatment of heart disease, cholesterol and its esters are of great interest. A separation of cholesterol from its esters and triglycerides is shown in Fig. 103. Another

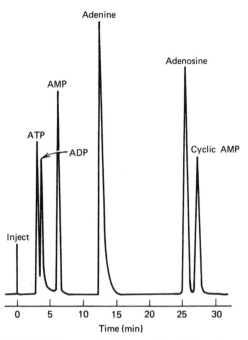

Figure 101 Analysis of adenine nucleotide standards. Sample: nucleotide standard mixture; column: μBondapak/C$_{18}$; gradient conditions: solvent: (A) 0.1 M NH$_4$H$_2$PO$_4$, pH 6.0; (B) CH$_3$OH/H$_2$O (80/20); 0 to 9% B, Curve 8, 20 min; flow rate: 2 ml/min; detector: model 440, 254 nm, 0.5 a.u.f.s. Reprinted with permission of Waters Associates Inc.

group of steroids, the aldosterones, are of interest in the control of hypertension and high blood pressure. Although aldosterone itself has a UV-absorbing chromophore, it has been difficult to analyze some of its metabolites because they do not absorb in the UV range. However, Morris et al. (103c) were able to develop an assay for aldosterone, 17-α-isoaldosterone, and the reduced metabolites of aldosterone, which they used in studies of aldosterone metabolism (Fig. 104).

7 VITAMINS

Although vitamins are not synthesized in the body, they are necessary for normal metabolic functions. Vitamins are a chemically diverse group

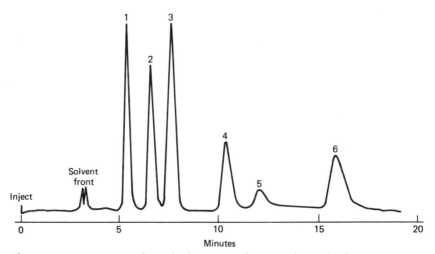

Figure 102 Separation of standard mixture of 6 steroids. Peak identities are: (1), estrone; (2), estradiol; (3), estradiol methyl ester; (4), 16-α-hydroxyestrone; (5), 16-oxo-estradiol and (6), estriol. Column, μBondapak/NH₂ (Waters); mobile phase, heptane/isopropanol (4:1, v/v); detection wavelength, 217 nm; temperature, ambient. Reproduced from reference 85 with permission.

of compounds and are usually divided into the broad classes of water-soluble and fat-soluble vitamins.

The water-soluble vitamins include vitamin C, lipoic acid, and the vitamin B complex, which consists of thiamine (B_1), riboflavin (B_2), nicotinic acid (niacin), pyridoxine (B_6), pantothenic acid (Coenzyme A), biotin, choline, inositol, *p*-aminobenzoic acid (the cogener of folic acid), and cyanocobalamin (B_{12}).

The vitamins of the B complex are frequently separated by cation exchange because of their basic properties (104–109); however, RPLC alone (110) or in combination with ion association (111) can also be used. By using fluorometric detection, the sensitivity can be improved over UV detection by at least an order of magnitude (112–114); Williams and Slavin (110) used an excitation wavelength of 453 nm and an emission filter of 520 nm to selectively detect riboflavin in the urine of normal subjects. A separation of three of the B_6 vitamins is shown in Fig. 105.

Although anion exchange has been used to determine vitamin C, both in body fluids and in pharmaceutical formulations (115,116), RPLC with tridecylammonium formate added to the mobile phase also produced excellent separations of ascorbic acid in multivitamin preparations and in food extracts (117). An example of the analysis of vitamin C in tomato juice is shown in Fig. 106. Since vitamin C does not have a strong chromophore, electrochemical detection is usually the method of choice (115,116).

The separation of vitamin A and its structurally similar isomers, such as the cis-trans retinals (120), has been classically performed by adsorption (120). Recently, separation of vitamin A and some of its isomers was accomplished by RPLC (118,119). Using RPLC, Zakaria et al. (121) separated the α and β carotenes, which are precursors of vitamin A, as shown in Fig. 107. Because of the extensive conjugation of the various vitamin A isomers, UV detectors give ample sensitivity. However, the

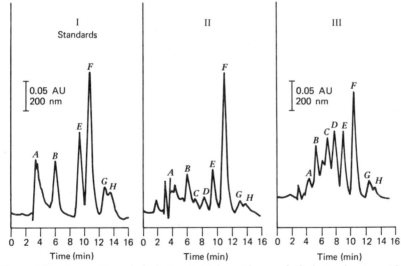

Figure 103 Separation of cholesterol, triglycerides, and cholesteryl esters. Chromatogram I represents a standard solution containing known quantities of the following materials in the 50μl injection cholesterol, 6μg (B); cholesteryl arachidonate, 3.5 μg (E); cholesteryl linoleate, 15 g (F); cholesteryl oleate, 5 μg (G); and cholesteryl palmitate, 2.5 μg (H). Chromatograms II and III represent 50 μl of 1:5 serum—IPA extracts. Peaks C and D represent unidentified glycerides and peak E represents unidentified glycerides and cholesteryl arachidonate. Chromatographic conditions: Column, μBondapak/C_{18} (30 × 4 mm I.D.); precolumn, C_{18} Porasil (5 cm × 1.2 mm I.D.); mobile phase, isopropanol-acetonitrile (50:50, v/v); flow rate, 1 ml/min. Reproduced from reference 103a with permission.

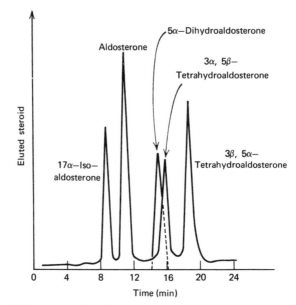

Figure 104 High pressure liquid chromatogram showing the separation of aldosterone, 17 α-iso-aldosterone, and the reduced metabolites of aldosterone. Column, μBondapak/C_{18} 0.4 × 30 cm; solvent, 50% aqueous mathanol; flow rate, 1 ml/min; temperature 25°C. Reproduced from reference 103c by courtesy of Marcel Dekker Inc.

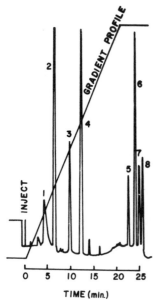

Figure 105 Simultaneous separation of water- and fat-soluble vitamins. peak identity: 1, vitamin B_6; 2, vitamin B_3; 3, vitamin B_1; 4, vitamin B_2; 5, retinol; 6, vitamin A acetate; 7, vitamin D_3; 8, vitamin E acetate. Chromatographic conditions: column, μBondapak/C_{18}; solvent, (A) 1% $(NH_4)_2CO_3$; (B) CH_3OH; gradient, linear from 0% B to 100% B in 20 min; detector: 254 nm, 0.16 a.u.f.s. Reproduced with permission of Waters Associates.

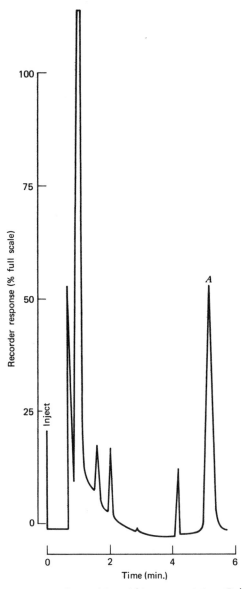

Figure 106 Chromatogram of ascorbic acid in tomato juice. Column, μBondapak/ C$_{18}$ (Waters) 30 cm × 4 mm i.d.; mobile phase, H$_2$O/tridecyl ammonium formate, (pH adjucted to 5.0) (50:50 v/v); flow rate, 25 ml/min. Reproduced from reference 133a with permission. Copyright (1976) American Chemical Society.

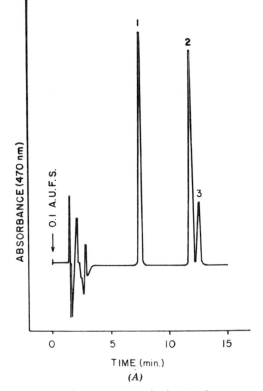

Figure 107 (A) Separation of carotene standards: #1, lycopene; #2, α-carotene; #3, β-carotene (R$_s$ for peaks #2 and 3, 1.46). Chromatographic conditions: column, Partisil-5/ODS, 5μm; eluent, 8.0% chloroform in acetonitrile; flow rate, 2.0 μl/min; temperature, ambient; detection, 470 nm; sensitivity, 0.1 a.u.f.s. (B) Chromatograms of a tomato extract: (a) 50 μl injected, (b) 10 μl injected; chromatographic conditions same as in (A). Reproduced from reference 121 with permission.

use of a variable wavelength US-vis detector (122) is advantageous since many of the vitamin A isomers have high molar absorptivities in the visible range. In addition, if the detector is equipped with the scanning capabilities, stopped-flow UV spectra of the separated peaks can be obtained and used for identification (121,123).

Vitamin E consists of a group of tocopherol isomers, and there have been many claims concerning the therapeutic value of large doses of α-tocopherol. Thus a rapid analysis of α-tocopherol was needed not only

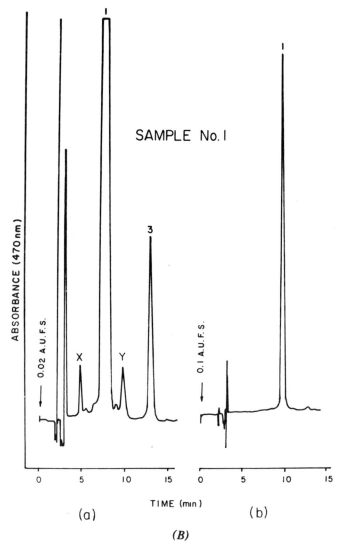

SAMPLE No. 1

(a)

(b)

(B)

Figure 107 (B) continued

in biomedical studies, but also in the pharmaceutical industry. Several HPLC methods of analysis for α-tocopherol in plasma and serum have been reported in the literature (124–127). Both adsorption on pellicular silica packings (128) and RP analyses on C_{18} packings (130) have been successfully used. Detection can be accomplished using a fixed-wavelength detector at 280 nm (129), a variable UV detector set at 292 nm (129), or a fluorescence detector. An example of a RPLC analysis of α-tocopherol using fluorescence detection with an excitation wavelength of 200 nm and an emission wavelength of 320 nm is shown in Fig. 108.

Vitamin D is found in nature as two isomers, D_2 and D_3 (calciferol and cholecalciferol), which are formed from the photochemical cleavage of the C_9-C_{10} bond of ergosterol and 7-dehydrocholesterol. Since the

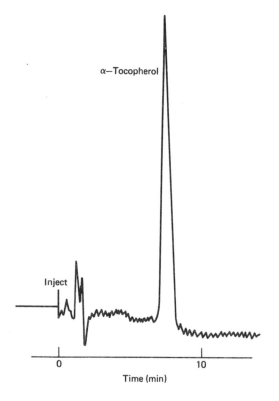

Figure 108   Liquid chromatographic analysis of α-tocopherol. Column, μBondapak/ C_{18}; solvent, CH_3OH/H_2O (95:5); flow rate, 2.0 ml/min; detector, fluorescence, 200 nm ex, 320 nm. Reprinted with permission of Waters Associates, Inc.

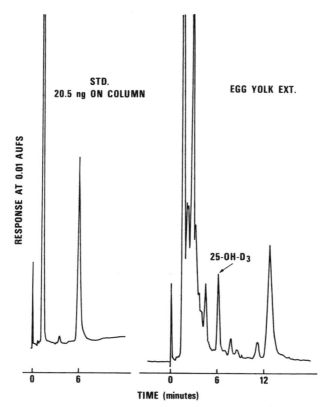

Figure 109 Liquid chromatograms showing 25-OH-D peaks from a standard and a chicken egg yolk extract. HPLC conditions: Zorbax ODS, 5 μm (DuPont) (2.1 mm × 25 cm); CH_3CN, CH_3OH, H_2O (94:3:3); 0.5 ml/min. Reproduced from reference 132 by courtesy of Marcel Dekker Inc.

metabolites of both of these vitamin D isomers show potential in the treatment of certain bone diseases (130), they are of great clinical importance. The predominant metabolites of vitamin D_2 and D_3 are the 25-hydroxy derivatives which were conventionally analyzed in body fluids by a competitive binding assay. However, interferences cause problems, and several sample preparation steps are required (132).

Although the majority of separations of vitamin D and its metabolites in serum or plasma have been performed by adsorption chromatography, the reversed phase has also been used. An example of a RPLC assay of the 25-OH-D_3 metabolite in egg yolk extract, as developed by Koshy and van der Slik (132), is shown in Fig. 109. Detection at 254 nm appeared

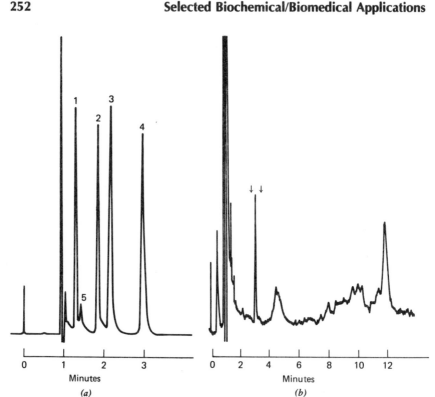

Figure 110 (a Separation of vitamin K compounds. Column, RP-18LL, (5 μm) (15 cm × 3.2 mm I.D.); mobile phase, 8% dichloromethane in acetonitrile, 1 ml/min; peaks, 1, MK-1; 2, MK-4; 3, K, epoxide; 5, unknown impurity. (b) Vitamin K in serum. Reproduced from reference 133 with permission.

to give adequate sensitivity for the serum concentrations normally encountered; however, because of the extremely sensitive detector setting needed (0.005 a.u.f.s), other detection methods should be investigated.

Vitamin K is of great clinical importance because of its role in the clotting of blood. Lefevre (133) reported on an assay for vitamin K which involves an extraction with hexane and fractionation on a silica column prior to the RP analysis. A chromatogram of standards of vitamin K and metabolites is shown in Fig. 110a and of vitamin K in serum in Fig. 110b.

8   MISCELLANEOUS

8.1   Creatinine

Creatinine, an excretion product of the kidney, occurs in muscle cells of vertebrates as the only intermediate product of creatine metabolism. The accurate measurement of endogenous creatinine is important in the monitoring of renal function and in the modification of therapeutic drug dosage for patients suffering from renal malfunctions. In addition, creatinine levels are being used as a basis for quantitative analysis of other urinary constituents, thus eliminating the variations associated with 24-hour urine collection. Creatinine concentrations in biological fluids are classically assayed by the Jaffé method (133b), although its nonspecificity is well recognized.

The increasing popularity of HPLC led to the development of several more specific analyses. Cation exchange was the first mode used to determine creatinine levels in plasma by HPLC (138a), and Soldin and Hill (134) developed a direct method for assaying creatinine by using ion-association chromatography and a RP microparticulate C_{18} column. Sodium lauryl sulfate was used as the counter ion, and deproteinated samples were injected directly into the chromatograph. Figure 111 shows a chromatogram of 6 μl of deproteinated serum. A total analysis time of 5 min was required. A serum concentration of 6 mg/l was readily detectable, using a wavelength of 200 nm. The creatinine analysis is an excellent example of the increased selectivity and accuracy that can be provided by RPLC for a clinical assay.

Creatinine levels in urine, anmiotic fluid, and CSF can also be assessed directly by RPLC with gradient elution and low wavelength (220 nm) UV detection (133). The analysis is rapid, and the sample preparation requires only filtration to remove the particulate matter. A representative chromatogram of a sample of amniotic fluid, analyzed for creatinine is shown in Figure 112.

8.2   Uric Acid

Uric acid, which is the end product of purine metabolism in humans, is routinely analyzed in the clinical laboratory by continuous methods, such

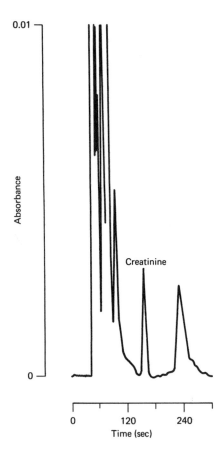

Figure 111 Ion-association separation of creatinine from other serum components using HPLC. Total serum volume used for the analysis, 10 μl; column, (Waters) μBondapak/C$_{18}$; mobile phase, methanol 80 mg/l sodium lauryl sulfate in 20 mmol/l phosphate buffer (pH 5.1) (76:24 v/v); flow rate, 2.3 ml/min; detection wavelength, 200 nm; temperature, 30°C. Reproduced from reference 134 with permission.

as the SMA 12/60 method, which involves dialysis, and the uricase method. While the enzymatic techniques (135) are more specific than those which involve chemical oxidation (136), interlaboratory variations still present some problems (137).

A RPLC method for the analysis of uric acid in serum has been introduced by Kiser et al. (138). This assay was designed for routine clinical use and as a reference method to check the results of continuous methods of analysis. Figure 113 (138) illustrates the separation of both standard solutions of uric acid and adenine as well as samples of serum. Excellent correlations were observed between the HPLC technique and the SMA 12/60 (dialysis) and uricase methods. While this newly developed HPLC method may not replace routine automated methods, it can be used as a readily available reference method to calibrate the routine methods against reference serum and to check for interferences.

9 THERAPEUTIC DRUGS AND DRUGS OF ABUSE

The use of HPLC to monitor the levels of therapeutic drugs and drugs of abuse in various tissues and body fluids has experienced a phenomenal growth over the past decade. Because of the "explosion" of applications in this area, any review is out of date almost before it can be published. Since an exhaustive survey of all areas of drug monitoring is not possible, only a few examples of recent developments will be discussed. However, there are some excellent reviews on therapeutic drug monitoring available in the literature (139–146).

Almost any drug that is administered therapeutically is subject to abuse

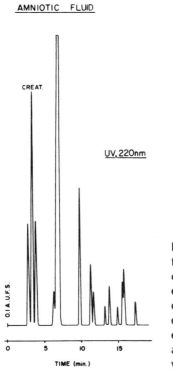

Figure 112 Creatinine in a sample of amniotic fluid. Volume injected, 100 μl. Chromatographic conditions: column, μBondapak/C_{18}; low strength eluent, 0.1 mol/l KH_2PO_4 (pH 2.5); high strength eluent mixture in CH_3CN and H_2O (3:2 v/v); gradient, linear from 0 to 80% of the high strength eluent in 45 min; flow rate, 1.4 ml/min; temperature, ambient. Reproduced from reference 133c with permission.

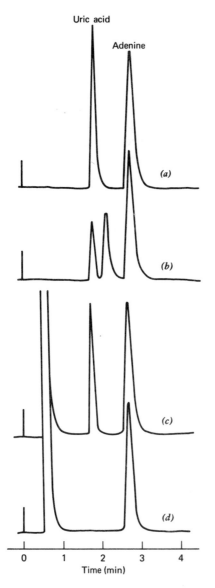

Figure 13 Uric acid in serum, measured by RPHPLC. (*A*) The separation of 100 mg/l aqueous uric acid solution, with adenine added as an internal standard; (*B*) a 100 mg/l aqueous uric acid standard prepared using formaldehyde, showing possible interferences that can be encountered when using such standards; (*C*) the analysis of Hyland reference serum (76 mg/l in uric acid); (*D*) the same serum pool after treatment with uricase, confirming the identity of the uric acid peak, and the absence of any compounds co-eluting with uric acid. Column, μBondapak/C$_{18}$; mobile phase, 35 ml of acetonitrile in 10 mmol/l phosphate buffer (pH 4.0); flow rate, 2.5 ml/min; detection wavelength, 280 nm; temperature, ambient. Reproduced from reference 138 with permission.

and thus can also be classified as a drug of abuse. The analytical methods used are the same for assaying therapeutic drugs and drugs of abuse; however, in the analysis of an illicit drug, the results must often stand up in a court of law and false positives can have criminal implications and serious penalties.

Comprehensive reviews (147–153) should be consulted for a detailed discussion on the analysis of common drugs of abuse, since we will only outline the role that HPLC can play in improving existing methods of analysis, particularly those which have classically been based on TLC and GC methods.

9.1 Antiarthritic Drugs

In the treatment of arthritis, salicylates, which have antiinflammatory properties, are widely used as the first line of treatment. Although the salicylates are relatively benign drugs with a rather wide latitude of safe dosages, it is important to be able to determine their plasma levels during long-term therapy. Noncompliance is a serious problem in arthritis therapy; thus physicians need to be able to determine if the plasma salicylate concentration is maintained at the level of 100 to 300 mg/l over the years of treatment (153a).

Many HPLC methods have been developed for the analysis of salicylates in tablet formulations (154), and Terweij-Groen and Kraak (155) used a liquid-liquid partition system to separate and identify aspirin and salicylic acid in human serum and urine.

Nonsteroidal, antiinflammatory drugs (NSAID) are now used widely in combination with aspirin for the treatment of rheumatoid arthritis. An example of a NSAID is indomethacin, which can be determined in physiological fluids by HPLC. However, most of the available HPLC methods do not measure both indomethacin and salicylate levels simultaneously (156). Recently, it was found to be possible to analyze both indomethacin and salicylic acid in serum using reversed phase and an isocratic elution (157). Naproxen [(+)-6-methoxy-α-methyl-2-naphthyleneacetic acid], another nonsteroidal drug used for the treatment of rheumatoid arthritis, can also be assayed, in the presence of its principle metabolites, using reversed phase (158,159,160a). Figure 114 (158) shows the separation of Naproxen from salicylic acid. The sensitivity of the analysis was greatly enhanced by the use of fluorescence detection and by prior extraction of the components of interest.

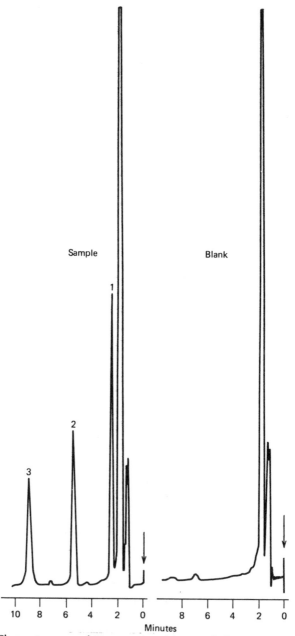

Figure 114 Chromatograms of Naproxen and its metabolite in urine. Chromatographic conditions: column, Lichrosorb RP-8 (100 × 4.4 mm); mobile phase, phosphate buffer (pH 7.0) and methanol (6/4, v/v); flow rate, 1.0 ml/min.; detection, fluorometric (excitation wavelength, 230 nm, no filter on the emission side); peak identity: 1,6-hydroxy-α-methyl-2-naphthylacetic acid (metabolite); 2,6-methoxy-2-naphthylacetic acid (internal standard); 3, Naproxen. Reproduced from reference 160a by courtesy of Marcel Dekker Inc.

Using the RP ion-association technique, the antiinflammatory drug DL-6-chloro-α-methylcarbazole-2-acetic acid has been recently assayed (160). A solution of 0.25% solution of tetrabutylammonium hydroxide in water methanol (30:70) was used as the mobile phase, and the compounds were detected fluorometrically. An ether extraction was used to prepare the plasma samples. The enhanced sensitivity of the fluorescent detector made possible the analysis of plasma concentrations of about 0.1 μmol/l, and the analysis time was under 10 min.

Another drug that is used in the treatment of gouty arthritis is the purine derivative allopurinol. In clinical assays of this drug, it is necessary to separate allopurinol not only from its metabolites, but also from the naturally occurring purines, so that accurate serum values of the drug can be determined. An example of the RPLC analysis of allopurinol is shown in Fig. 115.

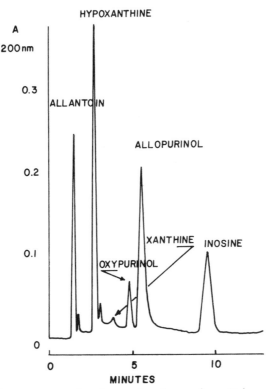

Figure 115 Chromatogram of purines and purine analogs. Column: Licrosorb RP-18, 10 μm (Knauer); mobile phase, 0.1 M phosphate buffer (pH 2.1); flow rate, 2.0 ml/min; temperature, 25°C. Reproduced with permission of Drs. I. Molnár and H. Knauer.

9.2 Antibiotics

Since the antibiotics are a large class of chemically diverse compounds, only a few examples of recent separations will be given. Due to the large number of antibiotics used clinically, this is an area where HPLC has enormous potential.

Penicillins have been separated using both ion exchange (161–164) and RPLC (165–169). An example of the RPLC separation of penicillin G and penicillin CK is shown in Fig. 116. Ion association was also used in the analysis of penicillin G (166). The cephalosporins have been separated on similar systems using both ion exchange (165,168,169) and RP (170).

There appears to be more HPLC separations (171–181) in the literature on the tetracyclines than on any other group of antibiotics. This is due, in part, to their ubiquity, good UV absorption, moderate polarity, and chemical stability. Pioneering work on these separations using RP systems was done by Knox et al. (174,175) (Fig. 117).

Another group of antibiotics of great interest is the sulfa drugs. A RPLC separation of 13 sulfonamides is shown in Fig. 118. Other anti-

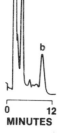

Figure 116   Antibiotics (penicillin G and penicillin VK) on Partisil-10 ODS. Operating conditions: 8; column, PXS-1025 ODS (4.6 mm × 250 nm); column temperature, ambient; mobile phase, 025 M NH$_4$NO$_3$/ 0025 M EDTA/MeOH; flow rate, 1.43 ml/min; pressure, 1000 psi; detection, UV at 254 nm; peaks 8, a), penicillin G; b), penicillin VK. Samples courtesy of: G = Squibb; VK =. Wyeth. Reprinted with permission of Whatman, Inc.

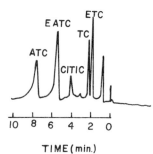

Figure 117 Chromatogram of tetracyclines on SAS-Hypersil showing near-optimum conditions with acetonitrile as modifier. Eluent, water-acetonitrile-acetic acid (71:18:5:10.5, v/v/v) containing $2.8 \cdot 10^{-3}$ M Na$_2$EDTA, 0.011 M KNO$_3$, and 0.06 M sodium acetate giving pH 3.0; detector, UV at 272 nm; 0.1 a.u.f.s. ATC, anhydrotetracycline; EATC, epianhydrotetracycline; CRTC, 7-chlorotetracycline; ETC, epitetracycline; TC, tetracycline. Reproduced from reference 175 with permission.

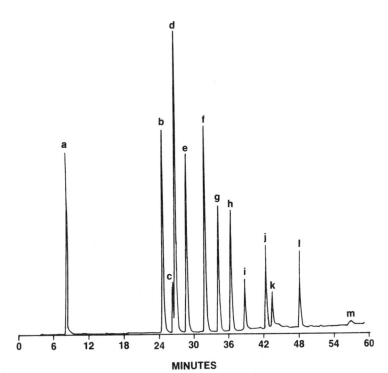

Figure 118 Separation of sulfonamides on Partisil-5 ODS. Operating conditions: 8; column, PXS-525 ODS (4.6 mm × 25 cm); column temperature, ambient; mobile phase, linear gradient 1.7% min, (1) water/methanol (90:10), (2) methanol/acetic acid (99:1); detection, UV at 254 nm. Peaks: a), sulfanilamide; b), sulfadiazine; c), sulfapyradine; e), sulfamerazine; f), sulfamethazine; g), sulfacholpyridazine; h), sulfasoxazole; i), sulfaethoxypyridazine; j), sulfadimethoxine; k), sulfaquinoxaline; l), sulfabromethazine; m), sulfaguanidine. Reprinted with permission of Whatman, Inc.

biotics that have been separated on RPLC systems include cerfuroxime (181), chloramphenicol (182), bacitracin (183,184), gentamicin (185,-186), and the polypeptide antibiotic gramicidin (187). The analysis of chloramphenicol in biological fluids is of great clinical interest due to its serious side effects at high concentrations (188). Thies and Fischer (182) developed a rapid analysis for therapeutic levels of this drug in 100 μl of serum extract, using a C_{18} microparticulate packing and direct detection at 278 nm.

9.3 Anticonvulsive Drugs

Since antiepileptic drugs are used over long periods of time, they are one of the most widely analyzed groups of drugs. Many HPLC assays have been reported (189–210). In the treatment of grand mal, petit mal, and psychomotor epilepsies, anticonvulsive drugs from several categories are often used in combination to effect therapy. Among the drugs that are widely used are phenobarbitol, primidone, ethosuximide, diphenylhydantoin, carbamazepine, and methsuximide. These drugs are sometimes metabolized to other equipotent forms; thus the development of assays for the drugs and their active metabolites can be a challenge to the clinical chemist. In the HPLC assays of anticonvulsive drugs, it is necessary either to extract the drugs from serum or urine or to deproteinate the sample. Extraction of the anticonvulsants may be accomplished by chloroform (196,201), ether (189), adsorption onto charcoal (189,208), or XAD-2 resin (209). However, deproteinization with acetonitrile (205) is advantageous not only because of its simplicity, but also because there are fewer interferences from endogenous compounds.

Absorption detectors are widely used for the anticonvulsive drugs. Detection at 254 nm is suitable for phenobarbitol and diphenylhydantoin, whereas nonspecific detection at 190 to 200 nm is needed for primidone, ethosuximide, methsuximide, and carbamazepine (189).

Although all the major modes of HPLC have been used for the anticonvulsant drugs, RP separation on a C_{18} column is now widely used. Figure 119 shows an isocratic RP separation of ethosuximide, primidone, phenobarbitol, diphenylhydantoin (DPH), methenytoin, and methsuximide standards. Miller and Tucker (Fig. 120) developed a method of determining diphenylhydantoin (dilantin) levels in serum using an ODS column and isocratic elution with a phosphate buffer. Compounds were detected by UV absorption at 220 nm.

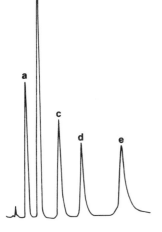

Figure 119 Separation of anticonvulsants. Operating conditions: 8; column, PSX-1025 ODS-2 (4.6 mm × 25 cm); column temperature, 45°C; mobile phase, acetonitrile/H_2O (17:83) flow rate, 2 ml/min; pressure, 1200 psi; detection, UV at 193 nm; peaks: (a), ethosuximide, (b), primidone, (c), phenobarbital, (d), mephenytoin, (e), methsuximide. Reprinted with permission of Whatman, Inc., and Micrometrics Instrument Corp.

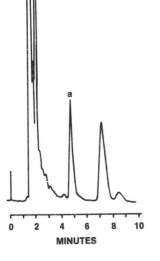

MINUTES

Figure 120 Drug screening of serum. Operating conditions: 8; column, PSX-1025 ODS (4.6 mm × 25 cm); column temperature ambient; mobile phase, MeOH/phosphate buffer (0.049 M in phosphoric acid and 0.49 M in KH_2PO_4, pH 2.3 before mixing) (40:60); flow rate, 2.0 ml/min; detection, UV at 220 nm; peaks: a), diphenylhydantoin (dilantin). Reprinted with permission of Whatman, Inc., Dr. J. A. Miller, Drew University, and Dr. E. Tucker, Morristown (N.J.) Hospital.

263

Carbamazepine is an interesting example of a drug whose metabolite is of equivalent potency to its cogener. Consequently, it is necessary to monitor not only the carbamazepine, but also its 10,11-epoxide. Mihaly and co-workers (201) used a chloroform extraction to separate the drug and its metabolites from the water-soluble compounds in plasma. The analysis was performed using a microparticulate C_{18} column with a methanol-water mobile phase (55/45, v/v) and UV detection at 254 nm. It was found that there was a large interindividual variation in the rates of metabolism of carbamazepine to its 10,11-epoxide. In addition, they found that there was a better correlation between the daily administered dose of carbamazepine and the plasma concentration of its 10,11-epoxide than with the plasma level of carbamazepine itself. This clearly illustrates the great utility of RPLC, since no other methods are available in which both the drug and its metabolites can be determined simultaneously on a routine basis. An example of the separation of carbamazepine and its metabolites is shown in Fig. 121.

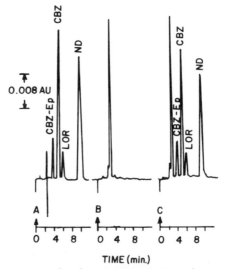

Figure 121 Chromatograms of carbamazepine (CBZ) and its 10, 11-epoxide (CBZ-EP). (A) Standard mixture of compounds, CBZ-EP, 0.625 μg; logazepam (LOR), 0.125 μg; N-desmethyldiazepam (ND), 0.375 μg; (B) Extract of drug-free plasma: (C) Standard mixture extracted from plasma. Chromatographic conditions: column, μBondapak/CN (10 μm); eluent, 30% mixture of acetonitrile in distilled-deionized water; flow rat, 1.2 ml/min; detection, 254 mm: temperature, ambient. Reproduced from reference 211 with permission.

9.4 Antidepressants

The tricyclic antidepressants are widely prescribed for the treatment of depression. This general class of drugs includes the dibenzocycloheptanes, dibenzocycloheptenes, dibenzazepines, dihydrodibenzazepines, and phenothiazines.

Although the tricyclic antidepressants are widely used, the optimal therapeutic concentrations in the plasma have not been established (211a–214). This is partly due to the variability in the response of similar patients to the same drug dosage (215–217); the effective plasma concentration is wide and is reported to be between 15 and 500 μg/l (218,219). Therefore, there is a critical need to obtain an accurate measurement of the levels of both the parent drug and its metabolites in plasma.

In addition, there is an increasing incidence of overdose characterized by epileptic seizures, possible cardiac failure, and eventual coma and death (220,221). Thus the clinician is faced both with the problem of monitoring therapeutic drug concentrations during therapy and of rapidly screening the patients' serum in the case of overdose.

Because of the presence of two benzochromophores, the tricyclics can easily be detected using an UV detector. Consequently, a number of HPLC assays for these compounds have been developed. In a recent review, HPLC methods were compared with other techniques for the benzodiazepines (222). Although adsorption columns have been used for the analysis of tricyclics, the partition mode (223–227) with a ternary-phase system (228) and RP ion association have been found to be effective. Knox and Jurand (229) have published a detailed study of the separation of some 20 of the tricyclic drugs, using both adsorption and RP ion association to produce the separation of 7 of the antidepressants, as shown in Fig. 122 (229).

9.5 Analgesic and Antipyretic Drugs

Analgesic and antipyretic drugs are perhaps the most widely used of all the drug classes. Because of their widespread use, not only for headaches and the common cold, but also in the treatment of more serious diseases like arthritis, gout, general neuralgia, and rheumatic fever, their analysis in plasma and other fluids is of considerable clinical importance.

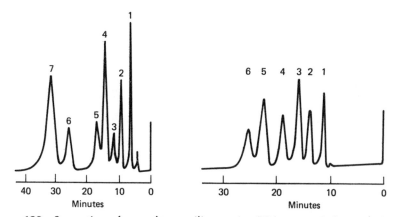

Figure 122   Separation of several tranquilizers using RP ion-association technique. (1), nortriptyline; (2), trimipramine; (3), propiomazine; (4), pipamazine; (5), thiopropazate; (6), perphenazine; (7), opipramol. Chromatogram on left: stationary phase, Merckosorb SI 100 (with 0.1 mol/l NaClO$_4$) loaded *in situ;* mobile phase, chloroform/L-butanol (30.70); linear velocity, 0.5 mm/sec; sample size, 1 μl, containing 100 to 300 ng of each solute. Chromatogram on the right: same conditions as above, but with methylene chloride/butanol/isoamyl alcohol (30:60:10, v/v/v) as in the mobile phase. Reproduced from reference 229 with permission.

The analgesic and antipyretic drugs fall into three broad chemical classes: the salicylates, *p*-aminophenol derivatives, and pyrazolon derivatives.

The salicylates include the esters of salicylic acid, in which the carboxylic acid group is esterified, and the salicylic acids, in which substitutions are made on the phenolic hydroxyl group. Methyl salicylate is an example of a carboxylate ester, whereas aspirin (acetylsalicylic acid) is an example of a phenolic ester. The *p*-aminophenol derivatives include drugs such as acetominophen, phenacetin, and acetanalid. The three most common pyrazolon derivatives are phenylbutazone, antipyrine, and aminopyrine. The first two categories of these drugs are widely sold without prescription, both singly and in various combinations. Acetominophen, for example, is sold either alone or in combination with other drugs such as phenacetin under more than 50 different brand names. Thus good analyses are vitally necessary for quality control in the pharmaceutical industry.

HPLC quickly proved itself to be uniquely suited for these highly polar compounds. One of the earliest separations of the analgesics by HPLC was the anion exchange method of Henry and Schmit (230). An example of an RPLC analysis of Excedrin is shown in Fig. 123.

The analysis of *N*-acetyl-*p*-aminophenol (Acetominophen, Paracetemol) in plasma and other body fluids is of clinical significance because of severe hepatic necrosis and renal failure resulting from overdose (231–234). Several RPLC procedures have been developed for the routine analysis of acetominophen and phenacetin in serum and in plasma (235–237). Detection has been accomplished either electrochemically (238) or, more commonly, by using a variable or fixed wavelength detector set at 250 to 254 nm (235,237). Extraction procedures are necessary for most of these analyses, and ether (236) and ethyl acetate (235) have been used for this purpose.

While the methods mentioned are useful for the rapid screening of overdose patients, it is often necessary to monitor not only the acetominophen and phenacetin blood levels, but also the concentrations of the various metabolites. In a comprehensive study of the metabolites of acetominophen in urine, Knox and Jurand (239) found that there are at least

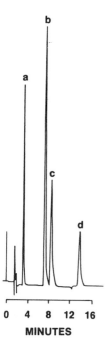

Figure 123 Analgesics (Excedrin) on Partisil-10 ODS. Operating conditions: 8; column, PXS-1025 ODS (4.6 mm × 25 cm); mobile phase, H_2O/MeOH (80:20) (pH 4.64) (H_3PO_4); flow rate, 1.50 ml/min; pressure, 1000 psi; detection, UV; peaks: (a), aspirin; (b), acetaminophen; (c), salicylamide; (d), caffeine. Reprinted with permission of Whatman, Inc.

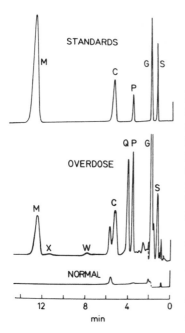

Figure 124 The separation of acetominophen (paracetamol) and its metabolites from the urine of overdose patients and from normal urine. Peak identities: (*M*), paracetamol mercapturic; (*C*), paracetamol cysteine; (*P*), paracetamol; (*G*), paracetamol glucuronide; (*S*), paracetamol sulphate. Peaks (*Q*), (*W*), and (*X*) are additional metabolites in the overdose sample. Chromatographic conditions: column, 12.5 cm × 5 mm I.D., packed with 6-μm silica to which octadecyltrichlorosilane had been chemically bonded. The packing material was further deactivated by exhaustive silanization. Eluent, water/isopropanol/formic acid (99:1:0.15, v/v/v); detection wavelength, 242 nm, 1.0 a.u.f.s. Reproduced from reference 239 with permission.

four major metabolites and that better RPLC separations were obtained when the residual silanol groups of the silica support were "capped" with trimethylsilyl groups. The separation of these principle metabolites in the urine of an overdose patient is shown in Fig. 124. MS was used as an ancilliary technique to aid in the identification of the compounds.

9.6 Bronchodialators

Theophylline and its derivatives are some of the most commonly used bronchodialators. The theophylline assay was one of the first routine HPLC analyses successfully used in the clinical laboratory. There is an optimal therapeutic range between 10 and 20 mg of theophylline per liter of plasma (240); however, toxic symptoms begin to occur when the plasma concentration exceeds 20 mg/l (241–243). Due to differences in metabolism, the actual dosage administered does not always correlate with the plasma concentration. Thus it is of the utmost importance that serum theophylline levels be accurately determined to ensure optimal clinical efficacy and safety.

HPLC has proven to be ideally suited for the analysis of the highly polar theophylline molecule and its related metabolites. While several excellent separations were obtained by ion-exchange chromatography (244,245), the use of RP packings has greatly reduced the separation time and increased the resolution of theophylline from its metabolites and other naturally occurring serum constituents (246–254). This is illustrated by the analysis of the theophylline standard and a serum sample from an asthmatic patient (Fig. 125).

One of the unique advantages of HPLC over the spectroscopic methods of analysis of theophylline is the ability of HPLC to monitor simultaneously both the theophylline and its many closely related metabolites. With RPLC, Desiraju et al. (255) were able to determine the half-lives and the metabolic fate of theophylline and its metabolites. Xanthine derivatives were added to serum, and the clearance of theophylline and its principal metabolites in urine was monitored over a 36-hour period.

9.7 Cardiovascular Drugs

Drugs that affect the cardiovascular system are chemically and pharmacologically diverse. They include the antiarrhythmia drugs, such as guinidine and procainamide, and the cardiac glycosides, such as digoxin, digitoxin, and lanatoside. The analyses of catecholamines and sympathomimetic drugs are discussed in another section.

There is great interest in the analysis of serum levels of cardioactive drugs, not only because they are widely used, but also because their toxicity is often significant, with a narrow margin of safety. Traditionally, the physician increases the dosage of the drugs until the onset of toxic symptoms and then reduces it to safer levels. This method leaves much to be desired, and a direct measurement of circulating drug levels is more effective and less toxic. Direct chemical analyses by classical spectrophotometric and fluorometric assays are usually nonspecific. In many cases, a circulating metabolite will have greater biological activity than its parent. Because of individual differences in metabolism, an assay that does not differentiate between the individual metabolites will often show poor dose-response linearity. Because of the tremendous number of drugs in this class, only a few examples will be given.

The cardioactive drugs, quinidine and procainamide, are effective against cardiac arrhythmias which are caused by abnormal impulse gen-

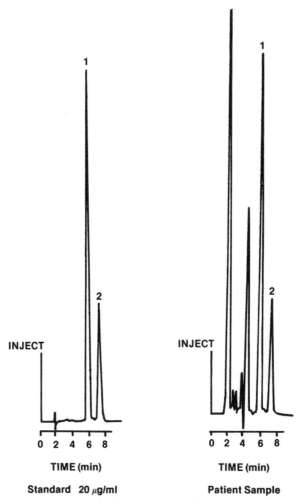

Figure 125 Analysis of theophylline in serum. Chromatographic conditions: column, μBondapak/C₁₈; solvent, 7/93 acetonitrile/0.01 *M* sodium acetate (pH 4.0); flow rate, 2 ml/min; detector, UV/visible absorbance detector 280 nm and/or 254 nm; chart speed, 0.2 in/min; sensitivity, 0.02 a.u.f.s. Reprinted with permission of Waters Associates, Inc.

eration. These drugs are used to return atrial fibrillation to normal sinus rhythm. Quinidine can be an especially dangerous drug because of the narrow range between effective therapeutic level and toxicity; toxic reactions begin at 10 mg/l of plasma, whereas the minumum effective therapeutic level is about 3 mg/l (256). Toxic symptoms include tinnitus, nausea, cinchonism, and atrial fibrillation, with possible death by cardiac failure (257).

HPLC methods have been highly successful in the analysis of both quinidine and its metabolites or contaminants and appear to be the methods of choice for the clinical chemist. Although both cation exchange (258) and the RP mode (259–262) have been successfully used, the latter technique is more commonly used for the analysis of quinidine and its metabolites. Crouthamel et al. (260) were able to separate quinidine and dihydroquinidine within 6 min, using a microparticulate C_{18} column. Detection was at 254 nm, with a reported sensitivity of about 0.1 mg/l of plasma. Powers and Sadee (262) were able to separate quinidine from several of its metabolites on an alkyl phenyl column, using isocratic elution. Chromatograms of samples prepared by two techniques are shown in Fig. 126.

Although quinidine and its metabolites fluoresce strongly, only one researcher has utilized fluorescence detection. Using an excitation wavelength of 340 nm and an emission wavelength of 418 nm, Dayer et al. (263) were able to achieve sensitivities of 0.5 mg/l of serum for quinidine and (3S)-3-hydroxyquinidine. The relative specificity of fluorescence should make this the preferred detection method for quinidine.

The effects of procainamide on the heart are almost identical to those of quinidine. Like quinidine, procainamide has a narrow therapeutic range (4 to 8 mg/l plasma). Toxicity is usually noticed when the plasma concentration exceeds 16 mg/l plasma (264). Procainamide is also similar to quinidine in that its major metabolite, N-acetylprocainamide, may have significant cardioactivity (264,265). Several excellent HPLC assays have been achieved using both normal phase (266) and RP (267,268) chromatography. The RP separation of procainamide from N-acetylderivatives of its metabolites is shown in Fig. 127.

Propranolol, an adrenergic β-receptor blocking drug, and phentolamine, an α-adrenergic blocking agent, have both been recently analyzed by HPLC (269,270). DeBros and Wolshin (270) used ion pairing with octane sulfonic acid on a RP column to assay for phentolamine, whereas Schmidt and Vandermark (269) used RPLC without ion association to

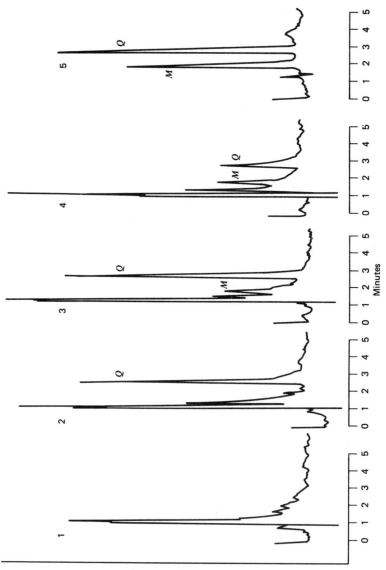

Figure 126 Separation of quinidine (Q) and a metabolite (M) from serum samples. (1), control serum; (2), quinidine (5 mg/l) added to control serum; (3), patient's serum with low metabolite level; (4), patient's serum with higher metabolite concentration; (5), same serum sample as No. (4), after alkaline extraction at pH 13. Column, microalkyl phenyl (Waters Associates); mobile phase, 40% (v/v) acetonitrile in 0.75 mol/l acetate buffer (pH 3.6); flow rate, 2.0 ml/min; detection at 330 nm, or at 236 nm, using a Schoeffel variable wavelength detector; temperature, ambient. Reproduced from reference 262 with permission.

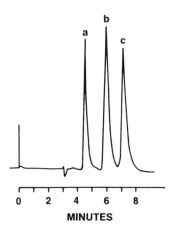

Figure 127 Cardiac drugs. Operating conditions: column, PXS-1025 ODS (4.6 mm × 25 cm); column temperature, ambient; mobile phase, MeOH/ phosphate buffer [0.049 M in phosphoric acid, 0.049 M in KH_2PO_4, pH 2.3 before mixing] (40:60); flow rate, 1.0 ml/min; detection, UV at 280 nm; peaks: (a), procainamide; (b), N-acetyl-procainamide; (c), N-propionylprocainamide (internal standard). Reprinted with permission of Whatman, Inc., Dr. J. A. Miller, Drew University, and Dr. E. Tucker, Morristown (N.J.) Hospital.

assay propranolol in plasma. By using fluorescent detection, they were able to detect as little as 500 pg of propranolol.

The introduction of the variable wavelength detector in HPLC has been an important contribution to the analysis of these compounds (271,272), since their absorption maxima are around 220 nm.

Since both adsorption and reversed phase have been used in assaying these drugs, Erni and Frei did a study on the complementary use of adsorption and RP chromatography for the analysis of both the cardiac glycosides and their aglycones in various pharmaceutical formulations (273). Since the compounds separated in a reversed elution order, it was found that the two modes are complementary and that selection of the appropriate column allows isolation of the compound of interest in minimum time.

9.8 Barbiturates and Other Drugs

As a class, the barbiturates ("goofballs, yellow jackets, red devil's") are probably one of the most widely used and abused drugs. Approximately 20 to 30 compounds in this class are in common use today. Since the barbiturates are weak acids with pK_a values in the range of 7.4 to 8.3, ion exchange has been used effectively (274,275). However, partition (277–280) and ion association (281) are now more widely used.

Although the individual members of the barbiturate class are similar in structure, they have distinct retention characteristics and can be readily separated from each other using RPLC (Fig. 128). Detection of the bar-

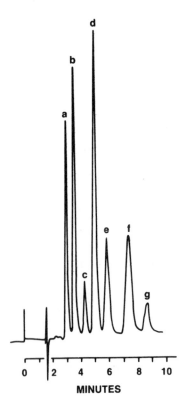

Figure 128 Barbiturates on Partisil-10 ODS. Operating conditions: column, PXS-1025 ODS (4.6 mm × 25 cm); column temperature, ambient; mobile phase, MeOH/phosphate buffer [0.049 M in phosphoric acid, 0.049 M in KH_2PO_4 (pH 2.3 before mixing)] (40:60); flow rate, 2.0 ml/min; detection, UV at 220 nm; peaks: (a), phenobarbital; (b), butabarbital; (c), mephobarbital; (d), pentabarbital; (e), secobarbital; (f), methohexital; (g), impurity. Reprinted with permission of Whatman, Inc., Dr. J. A. Miller, Drew University, and Dr. E. Tucker, Morristown (N.J.) Hospital.

biturates constitutes the most serious limitation of HPLC analyses. The wavelength of maximum absorption of the barbiturates, 205 nm, is too short for practical use. Clark and Chan (282) have overcome this problem in a simple yet elegant manner by post column alteration of the eluent pH. In the ionized form, that is, at a pH higher than 8, the λ_{max} of the barbiturates undergoes a bathochromic shift to about 245 nm. Therefore, a solvent pump was used to introduce a small, constant flow of borate buffer at pH 10 between the column outlet and the detector. Such postcolumn derivatization can be highly effective, and it has also been used for the amino acids, many of which have negligible UV absorption at practical wavelengths (283).

Narcotics, such as morphine, codeine, and heroin (284,285), have also been successfully analyzed by RPLC. For example, morphine can be separated from its metabolite, normorphine, as is shown in Fig. 129. This analysis can be carried out in a biological matrix such as rat brain.

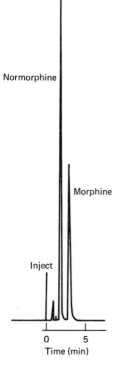

Normorphine

Morphine

Inject

0   5
Time (min)

Figure 129 Analysis of alkaloid standards. Sample, standard mixture, morphine/normorphine; column, μBondapak/C$_{18}$; solvent, 0.01 M (NH$_4$)$_2$HPO$_4$/CH$_3$ (70:30); flow rate, 2 ml/min; detector, UV: 254 nm, 0.2 a.u.f.s. Reprinted with permission of Waters Associates, Inc.

Figure 130 illustrates the analysis of morphine and normorphine, which had been added to the brain tissue.

Drugs of abuse can also be analyzed by RPLC. These drugs include LSD (286,287) and the cannabinoids (288–292). For other drugs of interest, there are many review articles as well as primary research articles that can be consulted.

10   FUTURE TRENDS

There is no doubt that in the near future RPLC will be used routinely in both clinical laboratories and biomedical research. Its use will be greatly facilitated by the use of microprocessors for data acquisition, instrument control, and data reduction, as well as for automatic sample injection, solvent control, and storage of data. The addition of microprocessing

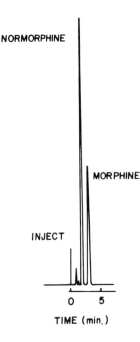

NORMORPHINE

MORPHINE

INJECT

0   5

TIME (min.)

Figure 130 Liquid chromatographic analysis of rat brain spiked with morphine/normorphine. Column, μBondapak/C$_{18}$; solvent, 0.01 M (NH$_4$)$_2$HPO$_4$/CH$_3$CN (70:30); flow rate, 2.0 ml/min; detector, UV: Model 440, 254 nm, 0.2 a.u.f.s. Reprinted with permission of Waters Associates, Inc.

systems will make RPLC more competitive with other methods now being used in the clinical laboratory.

RPLC will play several roles in the clinical laboratory; for example, for therapeutic drug monitoring, for detection and monitoring of disease states, and as a definitive method against which routine procedures can be checked. Thus RPLC may complement or replace some widely used methods such as radioimmunoassay and GC, especially where derivatization is required. RPLC also holds great potential for routine therapeutic drug monitoring in a doctor's office or in a small clinic, since patient compliance and correlation of plasma drug levels with prescribed dosage is a major problem in the treatment of chronic diseases. Thus compliance checks and plasma levels can be determined quickly and reliably without a delay in obtaining the results of the tests. In addition, RPLC can serve as a routine check on the automated methods and their calibration standards.

For trace analyses in which high efficiency columns are needed, short disposable cartridges (5 to 10 cm) of packing materials appear to be the columns of the future. Although they are used in Europe, manufacturers in the United States are just beginning to market them. With these cartridges, the sample dilution on the column is minimized. Furthermore, the installation of a new cartridge into the stainless steel column when resolution or efficiency deteriorates is easy and relatively inexpensive.

The development of packing materials with very small particle diameters (2 to 5 μm) promises to boost column efficiencies into the range of several hundred thousand plates per meter, although at the cost of increased operating pressures and possibly shorter column life. However, high efficiencies achieved with the ultrasmall particles may be needed in both the clinical laboratory and biomedical research for certain separations of complex mixtures of endogenous compounds, drugs, and their metabolites.

In addition, microbore and capillary columns capable of providing the desired information within seconds with extremely high efficiency will enhance further the utility of RPLC in the clinical laboratory.

The identification of separated solute bands, particularly in small amounts of sample containing trace levels of many compounds of interest, will undoubtedly receive considerable attention in the future. In all probability, multiple on-line detection systems will find increasing use in positive peak identification. Although at this time, the high cost, complexity, and need for skilled personnel limit the use of MS for routine work, the interfacing of HPLC with MS will eventually become the most powerful on-line combination for the elucidation of LC peak identities.

REFERENCES

Section 1

1. S. Moore, D. H. Spackman, and W. H. Stein, *Anal. Chem.*, **30**, 1185 (1958).
2. K. Ohtsuki, and H. Hatano, *J. Chromatogr.*, **148**, 536 (1978).
3. J. F. Riordan, and M. Sokolovsky, *Biochim. Biophys. Acta*, **236**, 161 (1971).
4. M. Sokolovsky, J. G. Riordan, and B. L. Vallee, *Biochem. Biophys. Res. Commun.*, **27**, 20 (1967).
5. T. Greibrokk, J. J. Hansen, R. Knudsen, Y. K. Lam, K. Folkers, and C. Y. Bowers, *Biochem. Biophys. Res. Commun.*, **67**, 338 (1975).
6. J. J. Hansen, T. Greibokk, B. L. Currie, K. N. G. Johansson, and K. Folkers, *J. Chromatogr.*, **135**, 155 (1977).
7. C. Horváth, W. Melander, and I. Molnár, *Anal. Chem.*, **49**, 142 (1977).

8.   C. Horváth, W. Melander, and I. Molnár, *J. Chromatogr.*, **125**, 129 (1976).

9.   B. Larsen, V. Viswanatha, S. Y. Chang, and V. J. Hruby, *J. Chromatogr.*, **149**, 241 (1978).

10.   E. Lundanes, and T. Greibrokk, *J. Chromatogr.*, **149**, 241 (1978).

11.   J. E. Rivier, *J. Liq. Chromatogr.*, **1**(3), 343 (1978).

12.   W. W. K. Fong and E. Grushka, *Anal. Chem.*, **50**, 1154 (1978).

13.   E. Grushka and R. P. W. Scott, *Anal. Chem.*, **45**, 1626 (1973).

13a.   E. J. Kikta, Jr. and E. Grushka, *J. Chromatogr.*, **135**, 367 (1977).

13b.   G. W. K. Fong and E. Grushka, *J. Chromatogr.*, **142**, 299 (1977).

14.   I. Molnár and C. Horváth, *J. Chromatogr.*, **142**, 623 (1977).

15.   W. S. Hancock, C. A. Bishop, R. L. Prestidge, and D. R. K. Harding, *Science*, **200**, 1168 (1978).

16.   F. Guyon, A. Foucault, and M. Caude, *J. Chromatogr.*, **186**, 677 (1979).

17.   W. T. W. Hearn, W. S. Hancock, and C. A. Bishop, *J. Chromatogr.*, **157**, 337 (1978).

18.   D. W. Hill, F. H. Walters, D. D. Wilson, and J. D. Stuart, *Anal Chem.*, **51**, 1338 (1979).

Section 2

19.   A. Carlsson, *Acta Neur. Scan.*, Suppl. No. 51, 11 (1972).

20.   G. G. Cotzias, P. S. Papvasiliov, and R. Gellene, *New Engl. J. Med.*, **280**, 337 (1969).

21.   U. S. von Euler, *Pharm. Rev.*, **6**, 15 (1954).

22.   S. S. Kety, *The Neurosciences, Second Study Program*, F. O. Schmitt, Ed., Rockefeller University Press, New York, 1970, p. 324.

23.   M. Jouvet, *Science*, **163**, 32 (1969).

24.   R. D. Myers and C. Chinn, *Am. J. Physiol.*, **224**, 230 (1973).

24a.   M. D. Armstrong, A. McMillan, and K. N. F. Shaw, *Biochim. Biophys. Acta*, **25**, 422 (1957).

24b.   T. T. Nestel and M. D. Esler, *Circ. Res.*, **26**, 75 (1970).

24c.   T. N. Mendel, D. L. Murphy, W. K. Engel, T. N. Chase, and E. Gordon, *Arch Neurol.*, **27**, 518 (1972).

24d.   S. E. Gitlow, L. M. Bertani, A. Rausen, D. Gribetz, and S. W. Dziedzic, *Cancer*, **25**, 1377 (1970).

25.   L. J. Marton, D. H. Russell, and C. C. Levy, *Clin. Chem.*, **19**, 923 (1973).

26.   F. Dreyfuss, R. Chayen, and G. Dreyfuss, *Isr. J. Med. S.*, **11**, 785 (1975).

27.   L. J. Marton, O. Heby, and V. A. Levin, *Cancer Res.*, **36**, 973 (1976).

28. P. T. Kissinger, R. M. Riggin, R. L. Alcorn, and L. D. Rau, *Biochem. Med.,* **13,** 299 (1975).

29. E. Ueda, N. Yoshida, K. Nishimura, T. Joh, K. Tsukada, S. Ganno, S. Antoku, and T. Kokubu, *Clin. Chim. Acta,* **80,** 447 (1977).

30. J. H. Knox and J. Jurand, *J. Chromatogr.,* **125,** 89 (1976).

31. L. D. Mell and A. B. Gustafson, *Clin. Chem.,* **24,** 23 (1978).

32. G. Schwedt and H. H. Bussemas, *Chromatographia,* **9,** 17 (1976).

33. J. H. Knox and J. Jurand, *J. Chromatogr.,* **125,** 89 (1976).

34. B. A. Persson and B. L. Karger, *J. Chromatogr.,* **122,** 305 (1976).

35. B. A. Persson and P. Q. Lagerstrom, *J. Chromatogr.,* **122,** 305 (1976).

36. I. Molnár and C. Horváth, *Clin. Chem.,* **22,** 1497 (1976).

37. R. M. Riggin and P. T. Kissinger, *Anal. Chem.,* **49,** 2109 (1977).

38. N. Seiler, B. Knodgen, and F. Eisenbeiss, *J. Chromatogr., Biomed. Appl.,* **145,** 29 (1978).

39. K. Imai, M. Tsukamoto, and Z. Tamura, *J. Chromatogr.,* **137,** 357 (1977).

40. K. Imai, *J. Chromatogr.,* **105,** 135 (1975).

41. K. Imai, M. Tsukamoto, and Z. Tamura, *J. Chromatogr.,* **137,** 357 (1977).

42. G. Schwedt, *J. Chromatogr.,* **143,** 463 (1977).

43. G. Schwedt and H. H. Bussemas, *Chromatographia.,* **9,** 17 (1976).

44. N. Seiler, *Methods Biochem. Anal.,* **18,** 259 (1970).

45. M. Roth, *Anal. Chem.,* **43,** 880 (1971).

46. E. Ueda, N. Yochida, K. Nishimura, T. Joh, K. Tsukada, S. Ganno, S. Antolu, and T. Kokubu, *Clin. Chim. Acta,* **80,** 447 (1977).

47. P. Haefelfinger, *J. Chromatogr.,* **48,** 184 (1970).

48. A. C. Moffat, E. C. Horning, S. B. Matin, and M. Rowalnd, *J. Chromatogr.,* **66,** 255 (1972).

49. A. M. Krstulović, M. Zakaria, K. Lohse, and L. B. Dziedzic, *J. Chromatogr.,* **186,** 733 (1979).

Section 3

49a. S. N. Pennington, *Anal. Chem.,* **43**(12), 1701 (1971).

50. J. Uberti, J. J. Lightbody, and R. M. Johnson, *Anal. Biochem.,* **80,** (1977).

51. J. A. Nelson, in *Advances in Chromatography,* Vol. 15, Calvin Giddings, E. Grushka, J. Cazes, and P. R. Brown, Eds., Dekker, New York, 1977, pp. 273–302.

52. R. A. Hartwick, A. Jeffries, A. M. Krstulović, and P. R. Brown, *J. Chromatogr. Sci.,* **16,** 427 (1978).

53. A. M. Krstulović, R. A. Hartwick, and P. R. Brown, *J. Chromatogr.*, **163**, 19 (1979).

Section 4

54. J. G. Turcotte, J. Y-K. Hsieh, and D. K. Welch., "General Method for the Analysis of Phosphoglycerides Using High Pressure Liquid Chromatography," Abstracts, 14th World Congress of the International Society for Fat Research, Brighton, England, September 1978.

55. J. Y-K. Hsieh, D. K. Welch, and J. G. Turcotte, "General Method for the Analysis of Phosphoglycerides Using High Pressure Liquid Chromatography," Ph.D. thesis, University of Rhode Island, 1979.

56. N. A. Porter, R. A. Wolf, and J. R. Nixon, "Separation and Purification of Lecithins by High Pressure Liquid Chromatography," *Lipids,* **14**, 20 (1979).

57. Technical notes, "Reverse Phases," from Brownlee Labs, Santa Clara, CA, Bulletin TN 910, 1977.

58. J. Y-K. Hsieh, D. K. Welch, and J. G. Turcotte, unpublished data.

59. N. E. Hoffman, and J. C. Liao, *Anal. Chem.*, **48**, 1104 (1976).

60. P. T-S. Pei, W. C. Kossa, and S. Ramachandran, *Lipids,* **11**, 814 (1976).

61. H. C. Jordi, *J. Liq. Chromatogr.*, **11**, 215 (1978).

62. R. A. Miller, N. E. Bussell, and C. Ricketts, *J. Liq. Chromatogr.*, **1**, 291 (1978).

63. R. H. McCluer, and F. B. Jungawala, *Adv. Exp. Med. Biolog.*, **68**, 533 (1976).

64. N. A. Porter, R. A. Wolf, and J. R. Nixon, *Lipids,* **14**, 20 (1979).

65. W. L. Erdahl and O. S. Privett, *Lipids,* **12**, 797 (1977).

66. S. Lam and E. Grushka, *J. Chromatogr.*, **158**, 207 (1978).

Section 5

67. M. Uziel, C. K. Koh and W. E. Cohn, *Anal. Biochem.*, **25**, 77 (1968).

68. C. A. Burtis, M. N. Munk, and F. R. MacDonald, *Clin. Chem.*, **16**, 667 (1970).

69. H. T. Bretter, G. Seibert, and R. K. Zahn, *J. Chromatogr.*, **140**, 251 (1977).

70. S. H.-Yu, P. P. L. Lau, and T. G. Spring, *Biochim. Biophys. Acta*, **517**, 31 (1978).

71. C. W. Gehrke, K. C. Kuo, G. E. Davis, R. D. Suits, T. P. Waalkes, and E. Borek, *J. Chromatogr.*, **150**, 455 (1978).

72. G. E. Davis, C. W. Gehrke, K. C. Kuo, and P. F. Agris, *J. Chromatogr.*, **173**, 281 (1979).

73. R. A. Hartwick, A. M. Krstulović, and P. R. Brown, *J. Chromatogr.*, **186**, 659 (1979).

74. A. M. Krstulović, R. A. Hartwick, and P. R. Brown, *Clin. Chim. Acta.,* **97,** 159 (1979).

75. R. A. Hartwick and P. R. Brown, *J. Chromatogr.,* **126,** 679 (1976).

76. N. E. Hoffman and T. C. Liao, *Anal. Chem.,* **49,** 2231 (1977).

77. R. A. Hartwick, S. P. Assenza, and P. R. Brown, *J. Chromatogr.,* **186,** 647 (1979).

78. R. S. Anderson and R. C. Murphy, *J. Chromatogr.,* **121,** 251 (1976).

79. A. M. Krstulović, P. R. Brown, and D. M. Rosie, *Anal. Chem.,* **49,** 2237 (1977).

80. P. R. Brown, A. M. Krstulović, and R. A. Hartwick, *J. Clin. Chem. Biochem.,* **14,** 282 (1976).

81. R. A. Hartwick and P. R. Brown, *J. Chromatogr. Biomed. Appl.,* **143,** 383 (1977)

82. A. M. Krstulović, R. A. Hartwick, P. R. Brown, and K. Lohse, *J. Chromatogr.,* **158,** 365 (1978).

83. S. Assenza and P. R. Brown, *J. Chromatogr. Biomed. Appl.,* **181,** 169 (1980).

84. S. Assenza and P. R. Brown, *Anal. Biochem.,* in press.

84a. P. R. Brown, A. M. Krstulovic, and R. A. Hartwick, *Human Hered.,* **27,** 167 (1977).

84b. P. D. Schweinsberg and T. L. Loo, *J. Chromatogr.,* **181,** 103 (1980).

Section 6

85. V. Fantl, C. K. Lim, and C. H. Gray, *High Pressure Liquid Chromatography in Clinical Chemistry,* P. F. Dixon, C. H. Gray, C. K. Lim, and M. S. Stoll, Eds., Academic, London, 1976, pp. 52–57.

86. F. A. Fitzpatrick, *Advances in Chromatography,* Vol. 16, J. C. Giddings, E. Grushka, J. Caazes and P. R. Brown, Eds., Dekker, New York, 1978, pp. 37–71.

87. R. Neher, *Advances in Chromatography,* Vol. 4, J. C. Giddings and R. A. Keller, Eds., Dekker, New York, 1967.

88. S. Siggia and R. Dishman, *Anal. Chem.,* **42,** 1223 (1970).

89. A. G. Butterfield, B. Lodge, N. J. Pound, and R. W. Sears, *J. Pharm. Sci.,* **64,** 441 (1975).

90. G. Cavina, G. Moretti, and A. Cantafora, *J. Chromatogr.,* **80,** 89 (1973).

91. F. A. Fitzpatrick, *Clin. Chem.,* **19,** 1293 (1973).

92. J. C. Touchstone, and W. Wortmann, *J. Chromatogr.,* **76,** 244 (1973).

93. D. E. Williamson, *J. Pharm. Sci.,* **65,** 138 (1976).

94. S. Siggia and R. Dishman, *Anal. Chem.,* **42,** 1223 (1970).

95. F. Bailey and P. N. Brittain, *J. Chromatogr.,* **83,** 431 (1973).

96. W. F. Beyer and D. D. Gleason, *J. Pharm. Sci.,* **64,** 1557 (1975).

97. F. A. Fitzpatrick, S. Siggia, and J. C. Dingman, *Anal. Chem.,* **44,** 2211 (1972).

98. G. Gordon and P. R. Wood, *Analyst,* **101,** 876 (1976).

99. R. E. Huetteman, and A. P. Shroff, *J. Chromatogr. Sci.*, **13**, 357 (1975).

100. M. Lafosse, G. Keravis, and M. H. Durand, *J. Chromatogr.*, **118**, 283 (1976).

101. K. Shimada, M. Hasegawa, K. Hasebe, Y. Jujii, and T. Nambara, *J. Chromatogr.*, **124**, 79 (1976).

102. J. W. Higgins, *J. Chromatogr.*, **121**, 329 (1976).

103. M. W. Gilgan, *J. Chromatogr.*, **129**, 447 (1976).

103a. I. W. Duncan, P. H. Culbreth, and C. A. Burtis, *J. Chromatogr. Biomed. Appl.*, **162**, 281 (1979).

103b. R. J. Dolphin and P. J. Pergande, *J. Chromatogr. Biomed. Appl.*, **143**, 267 (1977).

103c. D. J. Morris and R. Tsai, "Chromatographic Separation of Aldosterone and its Metabolites," in *Advances in Chromatography*, Vol. 19, J. Calvin Giddings et al., Eds., Dekker, New York, 1980.

Section 7

104. S. W. Bailey and J. E. Ayling, in *Chem. Biol. Pteridines*, Proc. Int. Symp., 5th, Pfleidere, (Ed.), W. De Gruyter, Berlin, 1975, pp. 6330–6643.

105. K. Callmer and L. Davies, *Chromatographia*, **7**, 644 (1974).

106. V. Noe and M. Psallidi, *Riv. Soc. Ital. Sci. Aliment.*, **5**, 137 (1977).

107. F. M. Rabel, *Chromatographia*, **8**, 156 (1975).

108. D. Whittmer and W. G. Haney, Jr., *J. Pharm. Sci.*, **63**, 588 (1974).

109. M. Wintrobe et al., Eds., *Principles of Internal Medicine*, 7th ed., McGraw-Hill, New York, 1974, pp. 1144–1171.

110. A. T. Williams and W. Slavin, *Chromatogr. Newslett.*, **5**, 9 (1977).

111. J. H. Knox and A. Yryde, *J. Chromatogr.*, **112**, 171 (1975).

112. D. R. Baker, R. C. Williams, and J. C. Steichen, *J. Chromatogr. Sci.*, **112**, 171 (1975).

113. J. J. Kirkland, *Analyst*, **99**, 859 (1974).

114. T. van de Weerdhof, M. L. Wiersum, and H. Reissenweber, *J. Chromatogr.*, **83**, 455 (1973).

115. P. T. Kissinger, L. J. Felice, R. M. Riggin, L. A. Pachla, and D. C. Wenke, *Clin. Chem.*, **20**, 992 (1974).

116. L. A. Pachla and P. T. Kissinger, *Anal. Chem.*, **48**, 364 (1976).

117. S. P. Sood, L. E. Sartork, D. P. Wittmer, and W. G. Haney, *Anal. Chem.*, **48**, 796 (1976).

118. C. V. Puglisi and J. A. F. DeSilva, *J. Chromatogr.*, **120**, 457 (1976).

119. R. C. Williams, J. A. Schmidt, and R. A. Henry, *J. Chromatogr. Sci.*, **10**, 494 (1972).

120. K. Tsukida, A. Kodama, and M. Ito, *J. Chromatogr.*, **134**, 331 (1977).

121. M. Zakaria, K. Simpson, P. R. Brown, and A. Krstulović, *J. Chromatogr.*, **176**, 109 (1979).

122. C. D. Carr, *Anal. Chem.*, **46**, 743 (1974).

123. R. W. Yost, *Chromatogr. Newslett.*, **5**, 44 (1977).

124. J. F. Cavins and G. E. Inglett, *Cereal Chem.*, **51**, 605 (1974).

125. T. Eriksson and B. Sorensen, *Act. Pharm. Scand.*, **14**, 475 (1977).

126. A. P. de Leeheer, V. O. De Bevere, A. A. Cruyl, and A. E. Claeys, *Clin. Chem.*, **24**, 585 (1978).

127. B. Nilsson, B. Johansson, L. Jansson, and L. Holmberg, *J. Chromatogr.*, **145**, 169 (1978).

128. B. Nilsson, B. Johansson, L. Jansson, and L. Holmberg, *J. Chromatogr.*, **145**, 169 (1978).

129. A. P. de Leenheer, V. O. DeBevere, A. A. Cruyl, and A. E. Claeys, *Clin. Chem.*, **24**, 585 (1978).

130. A. W. Norman, (Ed.), *Vitamin D and Problems Related to Uremic Bone Disease*, W. de Gruyter, Berlin, Germany, 1975.

131. R. S. Mason and S. Posen, *Clin. Chem.*, **23**, 806 (1977).

132. K. T. Koshy and A. L. van der Slik, *Anal. Lett.*, **10**, 523 (1977).

133. M. F. Lefevere, A. P. de Leenheer, and A. E. Claeys, *J. Chromatogr.*, **186**, 749 (1979).

133a. S. P. Sood, L. E. Sartori, P. P. Wittmer, and W. G. Haney, *Anal. Chem.*, **48**, 796 (1976).

Section 8

133b. M. Jaffé, *Z. Physiol. Chem.*, **10**, 391 (1886).

133c. A. M. Krstulovic, L. B. Dziedzic, and J. M. Caporusso, *Clin. Chim. Acta*, **99**, 189 (1979).

134. S. J. Soldin and J. G. Hill, *Clin. Chem.*, **24**, 747 (1978).

135. R. W. E. Watts, *Ann. Clin. Biochem.*, **11**, 103 (1974).

136. G. Lum and S. R. Gambino, *Clin. Chem.*, **24**, 536 (1978).

137. A. E. Rappoport, (Ed.), *Quality Control in the Clinical Laboratory*, Hans Huber, Bern, 1972, p. 304.

138. E. J. Kiser, G. F. Johnson, and D. L. Witte, *Clin. Chem.*, **24**, 536 (1978).

138a. W. L. Chiou, M. A. F. Gadalla, and E. W. Peng, *J. Pharm. Sci.*, **67**, 182 (1978).

Section 9

139. J. H. Done, J. H. Knox, and J. Loheac, *Application of High-Speed Liquid Chromatography*, Wiley, London, 1974.
140. B. B. Wheals and I. Jane, *Analyst,* **102,** 625 (1977).
141. R. E. Huetteman, M. L. Cotter, C. J. Shaw, and C. A. Janicki, *Anal. Chem.,* **47,** (5), 233R (1975).
142. R. G. Achari and E. E. Theimer, *J. Chromatogr. Sci.,* **15,** 320 (1977).
143. B. Fransson, K. G. Wahlund, I. M. Johansson, and G. Schill, *J. Chromatogr.,* **125,** 327 (1976).
144. P. J. Meffin, S. R. Harapat, Y. G. Yee, and D. C. Harrison, *J. Chromatogr.,* **138,** 183 (1977).
145. J. L. Powers and W. Sadee, *Clin. Chem.,* **24**(2), 299 (1978).
146. P. J. Watkins, *Clin. Chim. Acta.,* **18,** 191 (1967).
147. R. G. Achari and E. E. Theimer, *J. Chromatogr. Sci.,* **15,** 320 (1977).
148. A. Bye and M. E. Brown, *J. Chromatogr. Sci.,* **15,** 365 (1977).
149. A. T. Horváth and G. Clegg, *Clin. Chem.,* **24,** 804 (1978).
150. P. M. Kabra, H. Y. Koo, and L. J. Martin, *Clin. Chem.,* **24,** 657 (1978).
151. K. K. Kaistha, *J. Chromatogr.,* **141,** 145 (1977).
152. T. A. Rejent and K. C. Wahl, *Clin. Chem.,* **22,** 889 (1976).
153. H. Yatzidis, *Clin. Chem.,* **20,** 1131 (1974).

Section 9.1

153a. E. C. Boedeker and J. H. Dauber, Eds., *Manual of Medical Therapeutics*, 21st ed., Little, Brown, Boston, 1974.
154. R. P. Ascione and G. P. Chrekian, *J. Pharm. Sci.,* **64,** 1029 (1975).
155. C. P. Terweij-Groen, T. Vahlkamp, and J. C. Kraak, *J. Chromatogr.,* **145,** 115 (1978).
156. G. G. Skellern and E. G. Salole, *J. Chromatogr.,* **114,** 483 (1975).
157. M. McNeil and P. R. Brown, unpublished data.
158. R. F. Burgoyne, S. R. Kaplan, and P. R. Brown, *J. Liq. Chromatogr.,* **3**(1), 101 (1980).
159. D. Westerlund and A. Theodorsen, *J. Chromatogr.,* **144,** 27 (1977).
160. G. Palmskog and E. Hultman, *J. Chromatogr.,* **140,** 310 (1977).
160a. D. Westerlund, A. Theodorsen, and Y. Jaksch, *J. Liq. Chromatogr.,* **2**(7), 969 (1979).

Section 9.2

161. J. M. Blaha, A. M. Knevel, and S. L. Hem, *J. Pharm. Sci.*, **64**, 1384 (1975).

162. A. Bracey, *J. Pharm. Sci.*, **62**, 1695 (1973).

163. H. Bundgaard and C. Larsen, *J. Chromatogr.*, **132**, 51 (1977).

164. R. Saetre and D. L. Rabenstein, *Anal. Chem.*, **50**, 276 (1978).

165. V. Hartmann and M. Rodiger, *J. Chromatogr.*, **9**, 266 (1976).

166. J. H. Knox and A. Pryde, *J. Chromatogr.*, **112**, 171 (1975).

167. C. Larsen and H. Bundgaard, *J. Chromatogr.*, **147**, 143 (1978).

168. K. Tsuji and J. H. Robertson, *J. Pharm. Sci.*, **64**, 1542 (1975).

169. W. R. White, M. A. Carroll, J. E. Zarembo, and A. D. Bender, *J. Antibiot.*, **28**, 205 (1975).

170. R. D. Miller and N. Neuss, *J. Antibiot.*, **29**, 902 (1976).

171. F. Bailey, *Biochem. Soc. Trans.*, **3**, 861 (1975).

172. A. G. Butterfield, D. W. Hughes, W. L. Wilson, and N. J. Pound, *J. Pharm. Sci.*, **64**, 316 (1975).

173. G. Chevalier, C. Bollet, P. Rohrbach, C. Risse, M. Claude, and R. Rosset, *J. Chromatogr.*, **124**, 343 (1976).

174. J. H. Knox and J. Jurand, *J. Chromatogr.*, **110**, 103 (1975).

175. J. H. Knox and J. Jurand, *J. Chromatogr.*, **186**, 763 (1979).

176. A. P. de Leenheer and H. J. C. F. Nelis, *J. Chromatogr.*, **140**, 293 (1977).

177. R. F. Lindauer, D. M. Cohen, and K. P. Munnelly, *Anal. Chem.*, **48**, 1731 (1976).

178. E. Matusik and T. P. Givson, *Clin. Chem.*, **21**, 1899 (1975).

179. I. Nilsson-Ehle, T. T. Yoshikawa, M. C. Schotz, and L. B. Guze, *Antim. Ag. Ch.*, **9**, 754 (1976).

180. K. Tsuji and J. H. Robertson, *J. Pharm. Sci.*, **65**, 400 (1976).

181. K. Tsuji and J. H. Robertson, *J. Pharm. Sci.*, **65**, 400 (1976).

181a. K. Tsuji, J. H. Robertson, and W. F. Beyer, *Anal. Chem.*, **46**, 539 (1974).

182. R. L. Thies and L. J. Fischer, *Clin. Chem.*, **24**, 778 (1978).

183. K. Tsuji and J. H. Robertson, *J. Chromatogr.*, **112**, 663 (1975).

184. K. Tsuji, J. H. Robertson, and J. A. Bach, *J. Chromatogr.*, **99**, 597 (1974).

185. J. P. Anhalt, F. D. Sancilio, and T. Mc Corkle, *J. Chromatogr.*, **153**, 489 (1978).

186. G. W. Peng, M. A. Gadalla, A. Peng, W. Smith, and W. L. Chiou, *Clin. Chem.*, **23**, 1838 (1977).

187. K. S. Axelson and S. H. Vogelsang, *J. Chromatogr.*, **140**, 174 (1977).

188. A. A. Yunis and G. R. Bloomberg, *Prog. Hemat.*, **4**, 138 (1964).

Section 9.3

189. R. F. Adams, *Adv. Chromatogr.*, **15**, 131 (1977).
190. R. F. Adams and F. L. Vandermark, *High Pressure Liquid Chromatography in Clinical Chemistry*, P. F. Dixon et al., Eds., Academic, New York, 1976, pp. 143–153.
191. R. F. Adams and F. L. Vandermark, *Clin. Chem.*, **22**, 25 (1976).
192. R. F. Adams, G. J. Schmidt, and F. L. Vandermark, *Chromatogr. Newslett.*, **5**(1), 11 (1977).
193. S. H. Atwell, V. A. Green, and W. G. Haney, *J. Pharm. Sci.*, **64**, (1975).
194. A. Gugge, *J. Chromatogr.*, **128**, 111 (1976).
195. M. Eichelbaum and L. Bertilsson, *J. Chromatogr.*, **103**, 135 (1975).
196. F. J. Evans, *J. Chromatogr.*, **88**, 411 (1974).
197. G. Gauchel, F. D. Gauchel, and L. Birkofer, *Z. Klin. Chem. Klin, Biochem.*, **11**, 359 (1973).
198. K. Harzer and R. Barchet, *J. Chromatogr.*, **132**, 83 (1977).
199. I. M. House and D. J. Berru, *High Pressure Liquid Chromatography in Clinical Chemistry*, P. F. Dixon et al., Eds., Academic, New York, 1976, pp. 155–162.
200. P. M. Kabra and L. J. Martin, *Clin. Chem.*, **22**, 1070 (1976).
201. G. W. Mihaly, J. A. Phillips, W. J. Louis, and F. J. Vajda, *Clin. Chem.*, **23**(12), 2283 (1977).
202. R. J. Perchalski and B. J. Wilder, *Anal. Chem.*, **50**(4), 554 (1978).
203. R. W. Roos, *J. Pharm. Sci.*, **61**, 1979 (1972).
204. C. G. Scott and P. Bommer, *J. Chromatogr. Sci.*, **8**, 446 (1970).
205. S. J. Soldin and J. G. Hill, *Clin. Chem.*, **22**(6), 856 (1976).
206. T. B. Vree, B. Lenselink, E. van der Kleijn, and G. M. Nijhuis, *J. Chromatogr.*, **143**, 530 (1977).
207. J. M. Meola and M. Vanko, *Clin. Chem.*, **20**, 184 (1974).
208. J. M. Meola and M. Vanko, *Clin. Chem.*, **20**, 184 (1974).
209. M. Weinberger and C. Chidsey, *Clin. Chem.*, **21**, 834 (1975).
210. A. Frigerio and P. L. Morselli, *Advances in Neurology*, Vol. 11, J. K. Penry and D. C. Dely, Eds., Raven Press, New York, 1975, Chap. 16.
211. J. J. MacKichan, *J. Chromatogr., Biomed. Appl.*, **181**, 373 (1980).

Section 9.4

211a. M. Asberg, B. Chronholm, F. Sjoqvist, and D. Truck, *Br. Med. J.*, **iii**, 331 (1971).
212. R. O. Friedel and M. A. Raskind, *A Monograph of Recent Clinical Studies*, J. Mendels, Ed., Excerpta Medica, Princeton, NJ, 1975, pp. 51–53.

213. L. F. Gram, N. Reisby, and I. Ibsen, *Clin. Pharm.*, **19,** 795 (1976).

214. V. E. Ziegler, B. T. Co, and J. R. Taylor, *Clin. Pharm.*, **19,** 795 (1976).

215. J. T. Biggs, W. H. Holland, and W. R. Sherman, *Am. J. Psych.*, **132,** 960 (1975).

216. R. A. Braithwaite, R. Goulding, and G. Theano, *Lancet*, **i,** 1297 (1972).

217. A. H. Glassman and J. M. Perel, *Clin. Pharm.*, **16,** 198 (1974).

218. R. A. Braithwaite and B. Widdop, *Clin. Chim. Acta*, **25,** 461 (1971).

219. L. F. Gram and J. Christiansen, *Clin. Pharm.*, **17.** 555 (1975).

220. H. Gard, D. Knapp, and D. Hanenson, *Adv. Biochem. Psychopharm.*, **7,** 555 (1975).

221. D. G. Spiker, A. N. Weiss, and S. S. Chang, *Clin. Pharm.*, **18,** 539 (1975).

222. J. M. Clifford and W. F. Smyth, *Analyst*, **99,** 241 (1974).

223. K. Macek and V. Rehak, *J. Chromatogr.*, **105,** 182 (1975).

224. J. R. Salmon and P. R. Wood, *Analyst*, **101,** 611 (1976).

225. C. G. Scott and P. Bommer, *J. Chromatogr. Sci.*, **8,** 446 (1970).

226. P. J. Twitchett and A. C. Moffat, *J. Chromatogr.*, **111,** 149 (1975).

227. M. Viricel and M. Lemar, *J. Chromatogr.*, **116,** 343 (1976).

228. C. Gonnet and J. L. Rocca, *J. Chromatogr.*, **120,** 419 (1976).

229. J. H. Knox and J. Jurand, *J. Chromatogr.*, **103,** 311 (1975).

Section 9.5

230. R. A. Henry and J. Al Schmit, *Chromatographia*, **3,** 116 (1970).

231. R. G. Baum and F. F. Cantwell, *Anal. Chem.*, **50,** 280 (1978).

232. T. D. Boyer and S. L. Rouff, *J. Am. Med. A.*, **218,** 440 (1971).

233. B. McJunkin, K. W. Barwick, W. C. Little, and J. B. Winfield, *J. Am. Med. A.*, **23,** 1874 (1976).

234. J. Fuhr, *Arznei.-For.*, **14,** 74 (1964).

235. G. R. Gotelli, P. M. Kabra, and L. J. Marton, *Clin. Chem.*, **23,** 957 (1977).

236. R. A. Horvitz and P. I. Jatlow, *Clin. Chem.*, **23,** 1596 (1977).

237. T. G. Rosano, C. A. Brito, and J. M. Meola, *Chromatogr. Newslett.*, **6**(1), 1–4 (1978).

238. R. M. Riggin, A. L. Schmidt, and P. T. Kissinger, *J. Pharm. Sci.*, **64,** 2109 (1977).

239. J. H. Knox and J. Jurand, *J. Chromatogr.*, **142,** 651 (1977).

Section 9.6

240. K. M. Piafsky and R. I. Ogilvie, *N. Eng. J. Med.*, **292,** 1218 (1975).

241. P. M. Kabra, H. Y. Koo, and L. J. Martin, *Clin. Chem.*, **24,** 657 (1978).

242. G. Levy, in *Clinical Pharmacokinetics*, G. Levy, Ed., Am. Pharm. Ass., Washington, D.C. 1974, pp. 103–110.

243. M. W. Weinberger, R. A. Matthay, and E. J. Ginchansky, *J. Am. Med. A.*, **235**, 2110 (1976).

244. R. D. Thompson, H. T. Nagasawa, and J. W. Jeanne, *J. Lab. Clin. Med.*, **84**, 584 (1974).

245. M. Weinberger and C. Chidsey, *Clin. Chem.*, **21**, 834 (1975).

246. A. Osinga and F. A. DeWolff, *Clin. Chem. Acta*, **73**, 505 (1976).

247. M. A. Evenson and B. L. Warren, *Clin. Chem.*, **22**, 851 (1976).

248. L. C. Franconi, G. L. Hawk, B. J. Sandmann, and W. G. Haney, *Anal. Chem.*, **48**, 372 (1976).

249. W. J. Jusko and A. Poliszczuk, *Am. J. Hosp. P.*, **33**, 1193 (1976).

250. J. W. Nelson, A. L. Cordry, C. G. Aron, and R. A. Bartell, *Clin. Chem.*, **23**(1), 124 (1977).

251. J. J. Orcutt, P. P. Kozak, S. A. Gillman, and L. H. Cummins, *Clin. Chem.*, **23**, 599 (1977).

252. G. W. Peng, M. A. F. Gadalla, and W. L. Chiou, *Clin. Chem.*, **24**, 357 (1978).

253. G. W. Peng, V. Smith, A. Peng, and W. L. Chiou, *Res. Comm. Cp.*, **15**, 341 (1976).

254. O. H. Weddle and W. D. Mason, *J. Pharm. Sci.*, **65**, 865 (1976).

255. R. K. Desiraju and E. T. Sugita, *J. Chromatogr. Sci.*, **15**, 563 (1977).

Section 9.7

256. L. Goodman and A. Gilman, Eds. *The Pharmacological Basis of Therapeutics*, 3rd ed., Macmillan, New York, 1968, pp. 699–715.

257. M. Wintrobe et al., Ed., *Principles of Internal Medicine*, 7th ed., McGraw-Hill, New York, 1974, pp. 1144–1171.

258. R. E. Kates, D. W. McKennon, and T. J. Comstock, *J. Pharm. Sci.*, **67**(2), 269 (1978).

259. K. A. Conrad, B. L. Molk, and C. A. Chidsey, *Circulation*, **55**, (1977).

260. W. G. Crouthamel, B. Kowarski, and P. K. Narang, *Clin. Chem.*, **23**(11), 2030 (1977).

261. D. E. Drayer, K. Restivo, and M. M. Reidenberg, *J. Lab. Clin. Med.*, **90**(5), 816 (1977).

262. J. L. Powers and W. Sadee, *Clin. Chem.*, **24**(1), 299 (1978).

263. D. E. Drayer, K. Restivo, and M. M. Reidenberg, *J. Lab. Clin. Med.*, **90**(5), 816 (1977).

264. J. Elson, J. M. Strong, W. K. Lee, and A. J. Atkinson, Jr., *Clin. Pharm.*, **17**, 134 (1975).

265. D. E. Drayer, M. M. Reindenberg, and R. W. Sevy, *Proc. Soc. Esp. Biol. Med.*, **146**, 358 (1974).

266. J. S. Dutcher and J. M. Strong, *Clin. Chem.*, **23**(7), 1318 (1977).

267. K. Carr, R. L. Woosley, and J. A. Oates, *J. Chromatogr.*, **129**, 363 (1976).

268. R. M. Rocco, D. C. Ab, R. W. Giese, and B. L. Karger, *Clin. Chem.*, **23**(4), 705 (1977).

269. G. J. Schmidt and F. L. Vandermark, *Chromatogr. Newslett.*, **5**(3), 42 (1977).

270. F. deBros and E. M. Wolshin, *Anal. Chem.*, **50**(3), 521 (1978).

271. D. R. Baker, R. C. Williams, and J. C. Steichen, *J. Chromatogr. Sci.*, **12**, 499 (1974).

272. P. H. Cobb, *Analyst,* **101**, 768 (1976).

273. F. Erni and R. W. Frei, *J. Chromatogr.*, **130**, 169 (1977).

Section 9.8

274. M. Anders and J. Latoree, *Anal. Chem.*, **42**, 1430 (1970).

275. R. W. Roos, *J. Pharm. Sci.*, **61**, 1979 (1972).

276. J. E. Evans, *Anal. Chem.*, **45**, 2428 (1973).

277. B. Fransson, K. G. Wahlund, I. M. Johansson, and G. Schill, *J. Chromatogr.*, **125**, 327 (1976).

278. I. Halász, H. Schmidt, and P. Vogtel, *J. Chromatogr.*, **126**, 19 (1976).

279. U. R. Tjaden, J. C. Kraak, and J. F. K. Huber, *J. Chromatogr.*, **143**, 183 (1977).

280. P. J. Twitchett, A. E. P. Gorvin, and A. C. Moffat, *J. Chromatogr.*, **125**, 327 (1976).

281. B. Fransson, K. G. Wahlund, I. M. Johansson, and G. Schill, *J. Chromatogr.*, **125**, 327 (1976).

282. C. R. Clark and Jen-Lee Chan, *Anal. Chem.*, **50**, 635 (1978).

283. J. C. Kraask, K. M. Jonker, and J. F. K. Huber, *J. Chromatogr.*, **84**, 181 (1973).

284. I. Jane and B. B. Wheals, *J. Chromatogr.*, **84**, 181 (1973).

285. J. H. Knox and J. Jurand, *J. Chromatogr.*, **87**, 95 (1973).

286. J. Christie, M. W. White, and J. M. Wiles, *J. Chromatogr.*, **120**, 496 (1976).

287. R. O. Friedel and M. A. Raskind, *A Monograph of Recent Clinical Studies,* J. Mendels, Ed., Excerpta Medica, Princeton, NJ, 1975, pp. 51–53.

288. S. R. Abbott, *Res. Comm. Cp.*, **10**, 9 (1975).

289. E. E. Knaus, R. T. Coutts, and C. W. Kazakoff, *J. Chromatogr. Sci.*, **14**, 525 (1976).

290. R. N. Smith and C. G. Vaughan, *J. Chromatogr.*, **129**, 347 (1976).

291. B. B. Wheals, *J. Chromatogr.*, **122**, 85 (1976).

292. B. B. Wheals and R. N. Smith, *J. Chromatogr.*, **105**, 396 (1975).

Index

eration. These drugs are used to return atrial fibrillation to normal sinus rhythm. Quinidine can be an especially dangerous drug because of the narrow range between effective therapeutic level and toxicity; toxic reactions begin at 10 mg/l of plasma, whereas the minumum effective therapeutic level is about 3 mg/l (256). Toxic symptoms include tinnitus, nausea, cinchonism, and atrial fibrillation, with possible death by cardiac failure (257).

HPLC methods have been highly successful in the analysis of both quinidine and its metabolites or contaminants and appear to be the methods of choice for the clinical chemist. Although both cation exchange (258) and the RP mode (259–262) have been successfully used, the latter technique is more commonly used for the analysis of quinidine and its metabolites. Crouthamel et al. (260) were able to separate quinidine and dihydroquinidine within 6 min, using a microparticulate C_{18} column. Detection was at 254 nm, with a reported sensitivity of about 0.1 mg/l of plasma. Powers and Sadee (262) were able to separate quinidine from several of its metabolites on an alkyl phenyl column, using isocratic elution. Chromatograms of samples prepared by two techniques are shown in Fig. 126.

Although quinidine and its metabolites fluoresce strongly, only one researcher has utilized fluorescence detection. Using an excitation wavelength of 340 nm and an emission wavelength of 418 nm, Dayer et al. (263) were able to achieve sensitivities of 0.5 mg/l of serum for quinidine and (3S)-3-hydroxyquinidine. The relative specificity of fluorescence should make this the preferred detection method for quinidine.

The effects of procainamide on the heart are almost identical to those of quinidine. Like quinidine, procainamide has a narrow therapeutic range (4 to 8 mg/l plasma). Toxicity is usually noticed when the plasma concentration exceeds 16 mg/l plasma (264). Procainamide is also similar to quinidine in that its major metabolite, N-acetylprocainamide, may have significant cardioactivity (264,265). Several excellent HPLC assays have been achieved using both normal phase (266) and RP (267,268) chromatography. The RP separation of procainamide from N-acetylderivatives of its metabolites is shown in Fig. 127.

Propranolol, an adrenergic β-receptor blocking drug, and phentolamine, an α-adrenergic blocking agent, have both been recently analyzed by HPLC (269,270). DeBros and Wolshin (270) used ion pairing with octane sulfonic acid on a RP column to assay for phentolamine, whereas Schmidt and Vandermark (269) used RPLC without ion association to